Lecture Notes: Medical Microbiology and Infection

Lecture Notes

Medical Microbiology and Infection

Tom Elliott

BM BS BMedSci PhD DSc FRCPath
Consultant Medical Microbiologist & Director of Acute Medicine
University Hospital Birmingham NHS Foundation Trust, Birmingham

Tony Worthington

FIMBS CSci PhD
Lecturer in Clinical Microbiology
Aston University, Birmingham

Husam Osman

MB BCh PhD FRCPath
Consultant Virologist, Birmingham Public Health Laboratory
Birmingham Heartlands Hospital, Birmingham

Martin Gill

BSc MB ChB PhD FRCPath
Consultant Medical Microbiologist
University Hospital Birmingham NHS Foundation Trust, Birmingham

Fourth Edition

Blackwell Publishing

© 2007 Tom Elliott, Tony Worthington, Husam Osman, Martin Gill
Published by Blackwell Publishing Ltd

Blackwell Publishing, Inc., 350 Main Street, Malden, Massachusetts 02148-5020, USA
Blackwell Publishing Ltd, 9600 Garsington Road, Oxford OX4 2DQ, UK
Blackwell Publishing Asia Pty Ltd, 550 Swanston Street, Carlton, Victoria 3053, Australia

First published (*As Lecture Notes on Bacteriology*) 1967
First Edition 1975
Second Edition 1978
Reprinted 1979, 1983, 1986
Third Edition 1997
International edition 1997
Reprinted 2003, 2004
Fourth Edition 2007

Library of Congress Cataloging-in-Publication Data

Lecture notes. Medical microbiology & infection / Tom Elliott . . . [et al.].–4th ed.
    p. ; cm.
    Rev. ed. of: Lecture notes on medical microbiology / Tom Elliott, Mark Hastings, Ulrich Desselberger. 1997.
    Includes bibliographical references and indexes.
    ISBN-13: 978-1-4051-2932-9 (pbk. : alk. paper)
    ISBN-10: 1-4051-2932-8 (pbk. : alk. paper)   1. Medical microbiology–Outlines, syllabi, etc.
    [DNLM: 1. Microbiology.   2. Communicable Diseases–microbiology.   QW 4 L471 2006]
 I. Elliott, Tom, FRCPath.   II. Elliott, Tom, FRCPath. Lecture notes on medical microbiology & infection.

    QR46.G49 2006
    616.9'041–dc22

                                                                                          2006004743

ISBN-13: 978-1-4051-2932-9
ISBN-10: 1-4051-2932-8

A catalogue record for this title is available from the British Library

Set in 8/12 Stone Serif by SNP Best-set Typesetter Ltd., Hong Kong
Printed and bound in Singapore by Markono Print Media Pte Ltd

Commissioning Editor: Martin Sugden
Editorial Assistant: Ellie Bonnet
Development Editor: Mirjana Misina
Production Controller: Kate Charman

For further information on Blackwell Publishing, visit our website:
http://www.blackwellpublishing.com

Contents

Colour plates fall between pp. 120 and 121

Preface

The magnitude of recent changes in the field of medical microbiology has warranted this fourth edition of *Lecture Notes: Medical Microbiology and Infection*. While these changes have been encompassed in new chapters, this edition continues to maintain the well-received and user-friendly format of earlier editions, highlighting the pertinent key facts in medical microbiology and providing a sound foundation of knowledge which students can build on.

This fourth edition is now neatly arranged into four main sections: basic microbiology, applied microbiology, antimicrobial agents and infection. It covers all aspects of microbiology including bacteriology, virology, mycology and parasitology. New features in this edition include the ever-increasing problem of drug-resistant microorganisms, hospital-acquired infections and infections associated with the expanding use of medical devices. Furthermore, as in previous editions, the text is supported throughout with colour plates to illustrate the key points.

This book is written specifically for students in medicine, biomedicine, biology, dentistry, science and also pharmacology who have an interest in medical microbiology at both undergraduate and postgraduate levels. In addition, this book will serve as a useful *aide memoire* for doctors sitting FRCS and MRCP examinations as well as other healthcare professionals, for example biomedical scientists, working towards state registration.

Tom Elliott
Tony Worthington
Husam Osman
Martin Gill

Acknowledgements

We thank: Jennifer Leeming for comments on the mycology sections; Steve Smith for providing fluorescent stained fungi plates; Peter Lambert for helpful comments on the text; Lorraine Day for help and tolerance with preparing the manuscript; and Wyeth for an educational grant to support the publication of the colour plates.

Chapter 1

Basic bacteriology: structure

Bacterial structure

Bacteria are prokaryotic cells that are approximately 0.1–10.0 μm in size (Figure 1.1). They can exist in various shapes, including spheres (cocci), curves, spirals and rods (bacilli) (Figure 1.2); these form a basis for primary classification. Bacteria can be divided broadly into two main groups according to their Gram stain reaction which reflects the structure of their cell walls. Some bacteria stain Gram positive (blue/black) (Plates 3, 7, 11, 13, 17), whereas others are Gram negative (red) (Plates 18, 19). Bacterial structures (Figure 1.3) are described below.

Cell envelope

Cytoplasmic membrane

Cytoplasmic membranes surround the cytoplasm of all bacterial cells and are composed of protein, phospholipid and a small amount of carbohydrate; they resemble the membrane surrounding mammalian (eukaryotic) cells. The phospholipids form a bilayer into which proteins are embedded, some spanning the membrane. The membrane carries out many functions, including the synthesis and export of cell-wall components, respiration, secretion of extracellular enzymes and toxins, and the uptake of nutrients by active transport mechanisms.

Mesosomes are intracellular membrane structures, formed by an invagination of the cytoplasmic membrane. They more frequently occur in Gram-positive than in Gram-negative bacteria. Mesosomes present at the septum of Gram-positive bacteria are involved in chromosomal separation; at other sites they may be associated with cellular metabolism.

Cell wall

Bacteria maintain their shape by a strong rigid outer cover, the cell wall (see Figure 1.3).

Gram-positive bacteria have a relatively thick cell wall, largely composed of peptidoglycan, a complex molecule consisting of repeating sugar subunits cross-linked by peptide side chains (Figure 1.4a). Other cell-wall polymers, e.g. teichoic acid, are also present.

Gram-negative bacteria have a thinner peptidoglycan layer and an additional outer membrane that differs in structure from the cytoplasmic membrane (Figure 1.4b). The outer membrane contains lipopolysaccharides and, in some species, lipoprotein and porins, which are proteins involved in transport of substances across the cell envelope. Lipopolysaccharides are a characteristic feature of Gram-negative bacteria and are also termed 'endotoxins'. Endotoxins are released on cell lysis and have important biological activities involved in the pathogenesis of Gram-negative infections; they activate macrophages, clotting factors and complement, leading to disseminated intravascular coagulation and septic shock.

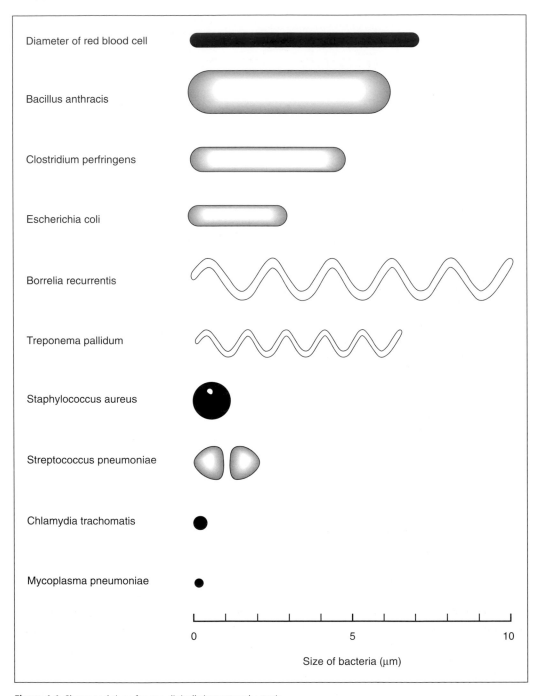

Figure 1.1 Shape and size of some clinically important bacteria.

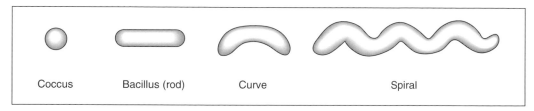

Figure 1.2 Some bacterial shapes.

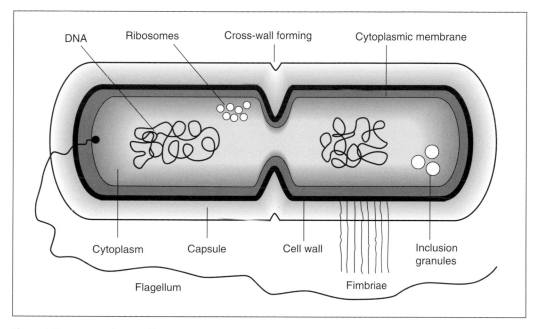

Figure 1.3 A section of a typical bacterial cell.

The cell wall is important in protecting bacteria against external osmotic pressure. Bacteria with damaged cell walls, e.g. after exposure to β-lactam antibiotics such as penicillin, often rupture (Plate 53). However, in an osmotically balanced medium, bacteria deficient in cell walls may survive in a spherical form called protoplasts. Under certain conditions some protoplasts can multiply and are referred to as L-forms. Some bacteria, e.g. mycoplasmas, have no cell wall at any stage in their life cycle.

The cell wall is involved in bacterial division. After the nuclear material has replicated and separated, a cell wall (septum) forms at the equator of the parent cell. The septum grows in, produces a cross-wall and eventually the daughter cells may separate. In many species the cells can remain attached, forming groups, e.g. staphylococci form clusters (Plate 3) and streptococci form long chains (Plate 7) (Figure 1.5).

Capsules

Some bacteria have capsules external to their cell walls (see Figure 1.3). These capsules are firmly bound to the bacterial cell and have a compact structure with a clearly defined boundary. They are often composed of high-molecular-weight polysaccharides.

The capsules are important virulence determinants in both Gram-positive and Gram-negative bacteria because they may protect the bacteria from host defences and, in some bacteria, aid attachment to host cells. Capsular antigens may also

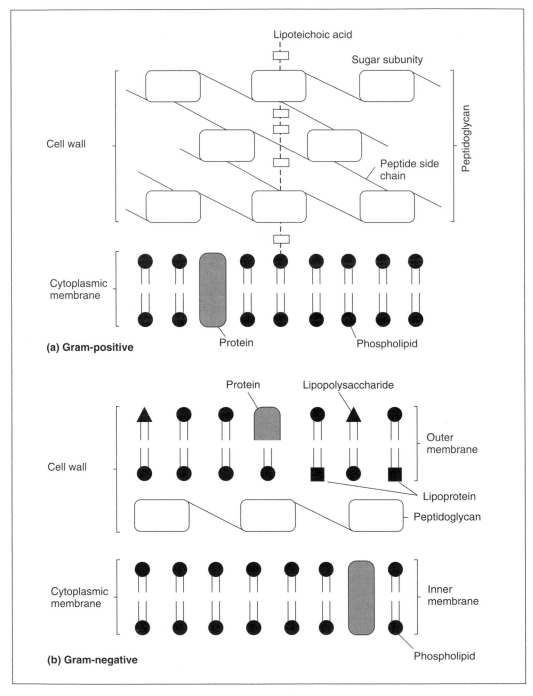

Figure 1.4 Cell wall and cytoplasmic membrane of (a) Gram-positive and (b) Gram-negative bacteria. The Gram-positive bacterial cell wall has a thick peptidoglycan layer with membrane-bound lipoteichoic acid. The Gram-negative bacterial cell wall has lipopolysaccharides in an outer membrane, with a thin inner peptidoglycan layer.

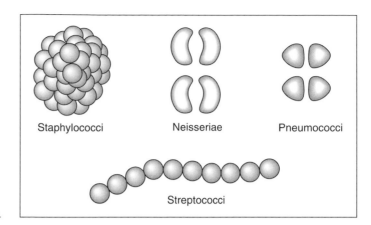

Figure 1.5 Some groups of bacteria.

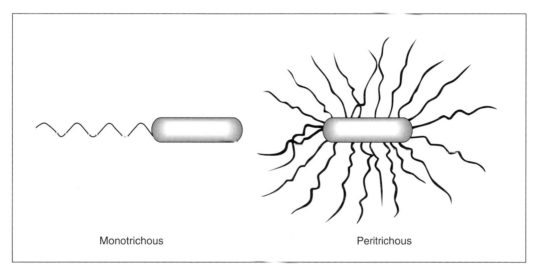

Figure 1.6 Arrangements of bacterial flagella.

be used in diagnosis to differentiate between isolates of the same bacteria, e.g. in the typing of *Streptococcus pneumoniae* for epidemiological purposes.

Bacterial slime

Extracellular slime layers are produced by some bacteria. They are more loosely bound to the cell surface than capsules and are also water soluble. The slime layer is composed predominantly of complex polysaccharides (glycocalyx) which may act as a virulence factor, e.g. in facilitating the attachment of *Staphylococcus epidermidis* on to artificial surfaces, such as intravascular cannulae (Plate 1) and prosthetic devices including artificial joints and heart valves.

Flagella

Bacterial flagella are 3–14 µm long and 0.02 µm in diameter (Plate 24); they are spiral-shaped filaments consisting mainly of the protein, flagellin. They can be single (monotrichous) or multiple (peritrichous) (Figure 1.6).

Flagella may facilitate locomotion in bacteria. They are observed under the light microscope with special stains. Flagella are usually detected for diagnostic purposes by observing motility in a bacterial suspension or by spreading growth on solid media. The antigenic nature of the flagella may be used to differentiate between and identify strains of salmonellae.

Fimbriae

Fimbriae or pili are thin, hair-like appendages on the surface of many Gram-negative, and some Gram-positive, bacteria (see Figure 1.3). They are approximately half the width of flagella, and are composed of proteins called pilins. In some bacteria they are distributed over the entire cell surface.

Fimbriae are virulence factors enabling bacteria to adhere to various mammalian cell surfaces, an important initial step in colonization of mucosal surfaces, e.g. *Neisseria gonorrhoeae* produce fimbriae that bind to specific receptors of cervical epithelial cells whereas *Streptococcus pyogenes* have fimbriae containing 'M' protein, which facilitates adhesion to human cells in the pharynx.

Specialized fimbriae are involved in genetic material transfer between bacteria, a process called conjugation.

Intracellular structures

Nuclear material

The bacterial nuclear material consists of a single circular molecule of double-stranded deoxyribonucleic acid (DNA), which is about 1 mm long when unfolded. It is tightly packed within the bacterium and is *not* surrounded by a nuclear membrane as in mammalian (eukaryotic) cells. Smaller extra-chromosomal DNA molecules, called plasmids, which can replicate independently, may also be present in bacteria. The chromosome usually codes for all the essential functions required by the cell; some plasmids control important phenotypic properties of pathogenic bacteria, including antibiotic resistance and toxin production. Extracellular nuclear material for encoding virulence and antibiotic resistance may also be transferred between bacteria and incorporated into the recipient's chromosome or plasmid. Transfer of genes encoding for virulence or antibiotic resistance may account for bacteria becoming resistant to antibiotics and for low-virulent bacteria becoming pathogenic.

Ribosomes

The cytoplasm has many ribosomes, which contain both ribonucleic acid (RNA) and proteins. Ribosomes are involved in protein synthesis.

Inclusion granules

Various cellular inclusions that serve as energy and nutrient reserves may be present in the bacterial cytoplasm. The size of these inclusions may increase in a favourable environment and decrease when conditions are adverse, e.g. *Corynebacterium diphtheriae* may contain high-energy phosphate reserves in inclusions termed 'volutin granules'.

Endospores

Endospores (spores) are small, metabolically dormant cells with a thick wall, formed intracellularly by members of the genera *Bacillus* and *Clostridium* (Plate 2). They are highly resistant to adverse environmental conditions and may survive desiccation, disinfectants or boiling water for several hours.

Spores are formed in response to limitations of nutrients by a complex process (sporulation) involving at least seven stages. When fully formed, they appear as oval or round cells within the vegetative cell. The location is variable, but is constant in any one bacterial species (Figure 1.7). Spores can

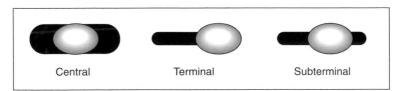

Central   Terminal   Subterminal

Figure 1.7 Size, shape and position of bacterial spores (from left to right): non-projecting, oval, central, e.g. *Bacillus anthracis*; projecting, spherical, terminal, e.g. *Clostridium tetani*; non-projecting, oval, subterminal, e.g. *C. perfringens*.

remain dormant for long periods of time. However, they are able to germinate relatively rapidly in response to certain conditions such as the presence of specific sugars.

Spores also have an important role in the epidemiology of certain human diseases such as anthrax, tetanus, gas gangrene and antibiotic-associated diarrhoea caused by *Clostridium difficile*.

The eradication of spores is also of particular importance in some processes, e.g. in the sterilization of instruments before surgery and in food canning, for the removal of *Clostridium botulinum* to prevent botulism.

Chapter 2

Basic bacteriology: physiology

Bacterial growth

Most bacteria will grow on artificial culture media. However, some bacteria, e.g. *Mycobacterium leprae* (leprosy) and *Treponema pallidum* (syphilis), cannot yet be grown in vitro; other bacteria, e.g. chlamydiae and rickettsiae, only replicate intracellularly within host cells and are therefore grown in tissue culture.

Under suitable conditions (nutrients, temperature and atmosphere) a bacterial cell will increase in size and then divide by binary fission into two identical cells. These two cells are able to grow and divide at the same rate as the parent cell, provided that conditions remain stable. This results in an exponential or logarithmic growth rate (Figure 2.1). The time required for the number of bacteria in a culture to double is called the generation time, e.g. *Escherichia coli* has a generation time of about 30 min under optimal conditions.

Requirements for bacterial growth

Most bacteria of medical importance require carbon, nitrogen, water, inorganic salts and a source of energy for growth. They have various gaseous, temperature and pH requirements, and can utilize a range of carbon, nitrogen and energy sources. Some bacteria also require special growth factors, including amino acids and vitamins.

Growth requirements are important in selecting the various culture media required in diagnostic microbiology and in understanding the tests for identifying bacteria.

Carbon and nitrogen sources

Bacteria are classified into two main groups according to the type of compounds that they can utilize as a carbon source.

1 *Autotrophs* utilize inorganic carbon from carbon dioxide and nitrogen from ammonia, nitrites and nitrates; they are of minor medical importance.

2 *Heterotrophs* require organic compounds as their major source of carbon and energy; they include most bacteria of medical importance.

Atmospheric conditions

Carbon dioxide

Bacteria require CO_2 for growth; adequate amounts are present in air or are produced during metabolism by the organisms themselves. A few bacteria, however, require additional CO_2 for growth, e.g. *Neisseria meningitidis, Campylobacter jejuni*.

Oxygen

Bacteria may be classified into four groups according to their O_2 requirements:

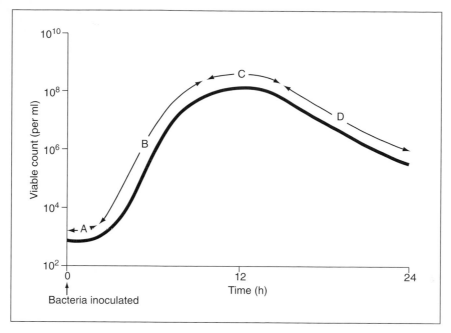

Figure 2.1 Bacterial growth curve showing the four phases: (A) lag; (B) log or exponential; (C) stationary; and (D) decline (death).

• *Obligate (strict) aerobes*: grow only in the presence of oxygen, e.g. *Pseudomonas aeruginosa*.
• *Microaerophilic bacteria*: grow best in low oxygen concentrations, e.g. *Campylobacter jejuni*.
• *Obligate (strict) anaerobes*: grow only in the absence of free oxygen, e.g. *Clostridium tetani*.
• *Facultative anaerobes*: grow in the presence or absence of oxygen, e.g. *Escherichia. coli*.

Temperature

Most pathogenic bacteria grow best at 37°C. However, the optimum temperature for growth is occasionally higher, e.g. for *C. jejuni*, it is 42°C. The ability of some bacteria to grow at low temperatures (0–4°C) is important in food microbiology; *Listeria monocytogenes*, a cause of food poisoning, will grow slowly at 4°C and has resulted in outbreaks of food poisoning associated with cook–chill products.

pH

Most pathogenic bacteria grow best at a slightly alkaline pH (pH 7.2–7.6). There are a few exceptions: *Lactobacillus acidophilus*, present in the vagina of postpubescent females, prefers an acid medium (pH 4.0). It produces lactic acid which keeps the vaginal secretions acid, thus preventing many pathogenic bacteria from establishing infection. *Vibrio cholerae*, the cause of cholera, prefers an alkaline environment (pH 8.5).

Growth in liquid media

When bacteria are added (inoculated) into a liquid growth medium, subsequent multiplication can be followed by determining the total number of viable organisms (viable counts) at various time intervals. The growth curve produced normally has four phases (see Figure 2.1):
1 Lag phase (A): the interval between inoculation of a fresh growth medium with bacteria and the commencement of growth.

2 Log phase (B): the phase of exponential growth; the growth medium becomes visibly turbid at approximately 10^4 cells/ml.

3 Stationary phase (C): the growth rate slows as nutrients become exhausted, waste products accumulate, and the rate of cell division equals the rate of death; the total viable count remains relatively constant.

4 Decline phase (D): the rate of bacterial division is slower than the rate of death, resulting in a decline in the total viable count.

Note that the production of waste products by bacteria, particularly CO_2, and the uptake of O_2 have been utilized in the development of modern semi-automated techniques to detect bacterial growth in blood cultures obtained from patients with septicaemia (Plate 47).

Growth on solid media

Liquid media can be solidified with agar that is extracted from sea weed. A temperature of 100°C is used to melt agar, which then remains liquid until the temperature falls to approximately 40°C, when it produces a transparent solid gel. Solid media are normally set in Petri dishes ('agar plates'). Most bacteria grow on solid media to produce visible colonies. Each colony comprises thousands of bacterial cells that emanated from either a single cell or a cluster of cells. The morphology of the bacterial colony assists in identification (Plate 4).

Growth on laboratory media

To culture bacteria in vitro, the microbiologist has to take into account the physiological requirements. Various types of liquid and solid media have been developed for the diagnostic microbiology laboratory (Plates 45, 48, 49).

Simple media

Many bacteria will grow in or on simple media, e.g. nutrient broth/nutrient agar that contains 'peptone' (polypeptides and amino acids from the enzymatic digestion of meat) and 'meat extract' (water-soluble components of meat containing mineral salts and vitamins).

Enriched media

These contain additional nutrients for the isolation of more fastidious bacteria that require special conditions for growth, e.g. agar containing whole blood (blood agar) (Plate 8) or agar containing lysed blood (chocolate agar).

Selective media

These are designed to facilitate growth of some bacteria, while suppressing the growth of others and include: *mannitol salt agar* which contains increased NaCl (salt) concentration for the recovery of staphylococci; and *MacConkey agar*, which contains bile salts and allows the growth of bile-tolerant bacteria only; and antibiotics, which are frequently added to media to allow only certain bacteria to grow while suppressing or killing others.

Indicator media

These are designed to aid the detection and recognition of particular pathogens. They are often based on sugar fermentation reactions that result in production of acid and the subsequent colour change of a pH indicator, e.g. MacConkey agar contains lactose and a pH indicator (neutral red); lactose-fermenting bacteria (e.g. *Escherichia coli*) produce acid and form pink colonies, whereas non-lactose fermenting bacteria (e.g. salmonellae) do not produce acid and form pale yellow colonies. This property facilitates the recognition of possible salmonella colonies among normal bowel flora. Note that indicator media may also contain selective agents including antibiotics or inhibitory substances such as bile salts. MacConkey agar is therefore both a selective medium and an indicator medium.

Chapter 3

Basic bacteriology: genetics

Replication of bacterial DNA

The normal mechanism of reproduction in bacteria is by binary fission. Genetic information is carried on double-stranded deoxyribonucleic acid (DNA), which, although arranged in a circular manner, is folded many times to fit inside the cell. An enzyme, DNA gyrase, is important in this folding process.

The structure and replication of bacterial DNA are similar to those of eukaryotic cells. Bacterial DNA consists of two complementary strands and, during cell division, these two strands separate with each acting as a template to allow the formation of two new double-stranded molecules.

Protein synthesis

DNA is transcribed into ribonucleic acid (RNA), which is in turn translated into new proteins via messenger RNA (mRNA) and ribosomes (Figure 3.1). The molecular basis of transcription and translation is important in the understanding of the action of some antibiotics, e.g. gentamicin, rifampicin and erythromycin, which inhibit protein synthesis.

The following are the main steps of protein synthesis:

1 *Transcription*: mRNA is transcribed from the chromosomal DNA strand:

2 *Translation*: codons (the nucleotide triplets) in mRNA specify the amino acid to be inserted into the forming polypeptide during translation; they are translated by binding of transfer RNA (tRNA) with the formation of a protein directed by a specific mRNA molecule that displays the appropriate corresponding nucleotides for each codon; the tRNA carries a matching amino acid:

- the ribosome binds to the mRNA
- an enzyme peptidyl transferase facilitates the transfer of each new amino acid to the growing peptide chain
- the mRNA is sequentially translated by this mechanism until a 'nonsense' codon is encountered that results in termination of the peptide chain.

Genotypic variation

The ability of bacteria to alter genetic information is fundamental to their survival in a changing environment. Such variation in the genome can occur in two ways: mutation and recombination (genetic transfer).

Mutation

During replication of DNA, copying errors called mutations may occur, leading to changes in the sequence of nucleic acids. Mutations can arise spontaneously or by exposure of bacteria to radiation or mutagenic chemicals.

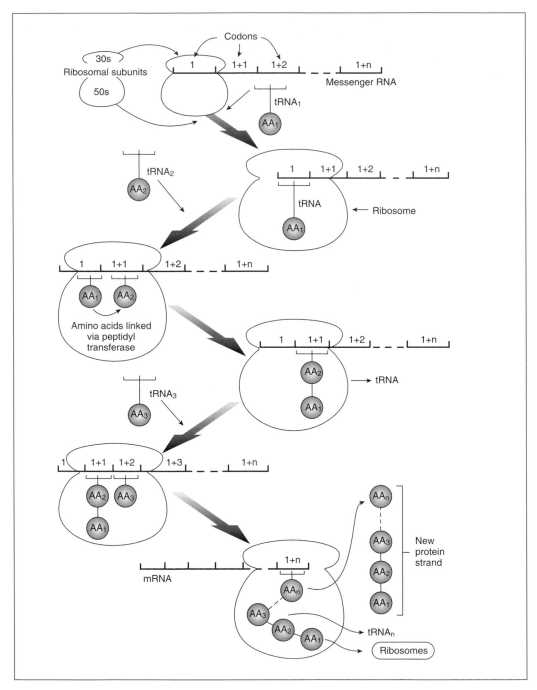

Figure 3.1 The major stages in protein synthesis involving ribosomal subunits (30S and 50S), messenger RNA (mRNA), transfer RNA (tRNA) and amino acid blocks (AA_n).

Many mutations result in a change in only a single nucleotide, with no detectable alteration in the end product, the transcribed protein. These are called point mutations.

More major changes, which lead to significant alterations in the organism, are often detrimental and the mutant organism may not survive. However, under certain circumstances, such alterations can result in a mutant cell with a significant advantage, allowing it to outgrow other daughter cells, e.g. antibiotic-resistant mutants may be selected out when that particular antibiotic is present in the environment. The principles of mutation and environmental selection are universal to biology.

Recombination (genetic transfer)

There are three basic mechanisms whereby new genetic information can transfer from one bacterium to another: transformation, conjugation and transduction (Figure 3.2).

Transformation

Some bacteria are able to take up soluble DNA fragments derived from other, normally closely related species, directly across their cell wall. This process of transformation was first described in *Streptococcus pneumoniae*.

Transduction

Viruses that infect bacteria are known as bacteriophages. During replication in bacteria, bacteriophages may incorporate some host DNA into their structures. When the virus is released and infects a new bacterial cell, the bacterial DNA from the donor cell may be integrated into the chromosome of the recipient. Bacteriophages infect only a narrow range of bacteria and thus this form of DNA recombination can occur only between closely related bacterial strains.

A bacteriophage can also incorporate its own viral DNA into the bacterial chromosome; occasionally, this can result in the bacteria synthesizing new proteins, e.g. diphtheria toxin and the erythrogenic toxin of group A β-haemolytic streptococci are both encoded on phage DNA.

Conjugation

Extrachromosomal segments of DNA, called *plasmids*, are present in some bacteria and are distinct from the chromosome. Plasmids replicate independently of the chromosome and, therefore, must encode certain essential genes that are necessary for DNA replication. Plasmids vary considerably in size, from large (80–120 kilobases [kb]) to small (2–10 kb).

Genes carried on plasmids may control one or more phenotypic characteristics of the bacterial cell, including antibiotic resistance and synthesis of bacterial toxins.

Certain plasmids, known as *conjugative plasmids*, are involved in the transfer of DNA between bacteria. Conjugative plasmids contain genes that control the formation of pili. These allow the bacterial cell to attach to a second cell via a cytoplasmic bridge. The conjugative plasmid divides and a copy is transferred across the cytoplasmic bridge into the recipient cell. Other non-conjugative plasmids can also be transferred in this way, provided that the host bacteria contains a conjugative plasmid to initiate the process.

The transfer of genetic information via plasmids can occur rapidly and between bacteria that are not closely related. Plasmid transfer of antibiotic-resistant genes is the most important mechanism by which resistance to antibiotics can spread. Some plasmids carry as many as six different antibiotic-resistant genes.

Movable genetic elements known as *transposons* ('jumping genes') are small genetic elements that can transfer between plasmids and chromosomal DNA within cells. At each end of the transposon there are specific base sequences known as insertion sequences that allow the transposon DNA to be inserted into existing DNA strands. Transposons allow genetic information to be transferred rapidly between plasmids and chromosomal DNA, facilitating the dissemination of genetic information among a bacterial population.

Genomic islands and horizontal gene transfer

With the advent of DNA sequencing a plethora of

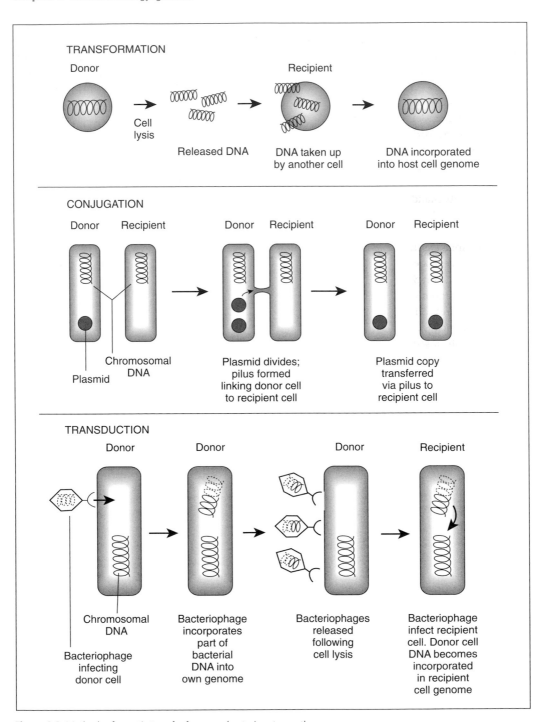

Figure 3.2 Methods of genetic transfer from one bacterium to another.

information in relation to the bacterial genome has become available. A recent finding is that the microbial genome can host significant amounts of foreign DNA incorporated through horizontal transfer of genetic material.

Foreign DNA incorporated into the host DNA is called a genomic island; however, depending on the function encoded by these islands, they may also be referred to as pathogenicity islands, symbiosis islands, metabolic islands or resistance islands.

Pathogenicity islands are the best understood genomic island and encode clusters of genes with products that contribute to virulence, e.g. adhesions, iron-uptake systems and toxins. Genomic islands may also encode for antibiotic resistance, e.g. methicillin-resistant *Staphylococcus aureus* (MRSA) contains a genomic region known as *mecA* gene, the expression of which results in resistance to methicillin. *mecA* is located on a large chromosomally inserted element called SCCmec (staphylococcal cassette chromosome mec).

MRSA may have originally acquired its resistance to methicillin after transfer of genetic material from a methicillin-resistant coagulase-negative staphylococcus.

Horizontal gene transfer of genomic islands between microorganisms is therefore a powerful mechanism by which the bacterial genome can be altered.

Genetic engineering

The various vectors that allow the transfer of DNA between bacteria form the basis of genetic engineering. Genes coding for important products can be inserted into plasmids or bacteriophages and used to transfect other organisms of the same or different species, allowing the synthesis of large amounts of the defined product. These techniques have been utilized in the field of vaccine research and production, and for studying bacterial pathogenicity and resistance to antimicrobial agents.

Chapter 4

Classification of bacteria

Bacterial taxonomy and nomenclature

The classification of microorganisms is essential for the understanding of clinical microbiology. Bacteria are designated by a binomial system, with the genus name (capital letter) followed by the species name (without capital letter), e.g. *Escherichia coli* or *Staphylococcus aureus*. Names are often abbreviated, e.g. *E. coli* and *S. aureus*.

Many nomenclature problems exist with this system that can lead to confusion, e.g. 'bacillus' refers to any rod-shaped bacteria, whereas the genus *Bacillus* includes only the aerobic spore-bearing rods. Other complications include the use of alternative terminology. *Streptococcus pneumoniae*, is referred to as the pneumococcus and *Neisseria meningitidis* as the meningococcus. Occasionally, collective terms are used, e.g. the term 'coliform' may indicate *E. coli* or a closely related Gram-negative bacillus found within the gut, and the term 'coagulase-negative staphylococci' means staphylococci other than *S. aureus*. In this text, conventional terminology is used and, where appropriate, common alternatives are indicated.

Bacterial classification

Medically important bacteria can be subdivided into five main groups according to their morphology and staining reactions. The basic shape of bac-teria include cocci, bacilli, and spiral and pleomorphic forms. Each of these morphological forms is further subdivided by their staining reactions, predominantly the Gram and acid-fast stains (Table 4.1). The more important pathogenic bacteria are divided primarily into Gram-positive (Plate 7) or Gram-negative (Plate 19) organisms. Other characteristics, including the ability to grow in the presence (aerobic) or absence (anaerobic) of oxygen and spore formation, are used to divide the groups further. Subdivision of these groups into genera is made on the basis of various factors, including culture properties (e.g. conditions required for growth and colonial morphology), antigenic properties and biochemical reactions. The medically important genera based on this classification are shown in Table 4.2 (Gram positives) and Table 4.3 (Gram negatives).

Table 4.1 Main groups of pathogenic bacteria.

PATHOGENIC BACTERIA
I Gram-positive cocci, bacilli and branching bacteria
II Gram-negative cocci, bacilli and comma-shaped bacteria
III Spiral-shaped bacteria
IV Acid-fast bacteria
V Cell-wall-deficient bacteria

Table 4.2 Classification of Gram-positive bacterial pathogens.

GRAM-POSITIVE BACTERIA			
Grouping	**Aerobic/anaerobic growth**	**Genus**	**Examples of clinically important species**
Gram-positive cocci			
Clusters	Both	*Staphylococcus*	*S. aureus, S. epidermidis, S. saprophyticus*
Chains/pairs	Both	*Streptococcus*	*S. pneumoniae, S. pyogenes, Enterococcus faecalis*
Squares	Both	*Micrococcus*	
Chains	Anaerobic	*Peptococcus* and *Peptostreptococcus*	
Gram-positive bacilli			
Sporing	Aerobic	*Bacillus*	*B. anthracis, B. cereus*
Non-sporing	Both	*Corynebacterium*	*C. diphtheriae*
	Aerobic or microaerophilic	*Listeria*	*L. monocytogenes*
	Anaerobic or microaerophilic	*Lactobacillus*	
Sporing	Anaerobic	*Clostridium*	*C. difficile, C. botulinum, C. perfringens, C. tetani*
Non-sporing	Anaerobic	*Propionibacterium*	*P. acnes*
Branching	Anaerobic	*Actinomyces*	*A. israeli*
	Aerobic	*Nocardia*	*N. asteroides*

Table 4.3 Classification of Gram-negative bacterial pathogens.

GRAM-NEGATIVE BACTERIA				
Shape	**Aerobic/ anaerobic growth**	**Major grouping**	**Genus**	**Examples of clinically important species**
Cocci	Aerobic		*Neisseria*	*N. gonorrhoeae* *N. meningitidis*
Cocci	Anaerobic		*Veillonella*	
Bacilli		Enterobacteriaceae ('Coliforms')	*Enterobacter*	*E. chloaceae*
			Escherichia	*E. coli*
			Klebsiella	*K. pneumoniae*
			Proteus	*P. mirabilis*
			Salmonella	*S. typhimurium*
			Serratia	*S. marcescens*
			Shigella	*S. sonnei*
			Yersinia	*Y. enterocolitica*
Bacilli	Aerobic		*Pseudomonas*	*P. aeruginosa*
Comma shaped	Both	Vibrios	*Vibrio*	*V. parahaemolyticus* *V. cholerae*
			Campylobacter	*C. jejuni*
			Helicobacter	*H. pylori*
Bacilli	Varies with genus	Parvobacteria	*Bordetella*	*B. pertussis*
			Brucella	*B. abortus*
			Haemophilus	*H. influenzae* *H. parainfluenzae*
			Pasteurella	*P. multocida*
Bacilli	Aerobic		*Legionella*	*L. pneumophila*
Bacilli	Anaerobic		*Bacteroides*	*B. fragilis*
			Fusobacterium	

Other bacterial groups

Spiral bacteria

These are relatively slender spiral-shaped filaments, which are classified into three clinically important genera:

1 *Borrelia*: these are relatively large, motile spirochaetes and include *Borrelia vincenti* and *Leptotrichia buccalis*, which cause Vincent's angina, and *B. recurrentis*, which causes relapsing fever.

2 *Treponema*: these are thinner and more tightly spiralled than *Borrelia*. Examples include *Treponema pallidum* (causes syphilis) and *T. pertenue* (causes yaws).

3 *Leptospira*: these are finer and even more tightly coiled than the treponemes. They are classified within the single species of *Leptospira interrogans*, which is divided serologically into two complexes. There are over 130 serotypes in the interrogans complex, many of which are pathogenic, including *L. icterohaemorrhagiae* (causes Weil's disease) and *L. canicola* (causes lymphocytic meningitis).

Acid-fast bacilli

These include the genus *Mycobacterium*. They are identified by their acid-fast staining reaction (Plate 26), which reflects their ability to resist, after being stained with hot carbol fuchsin, decolorization with acid. Mycobacteria are generally difficult to stain by Gram's method. They can be simply divided into the following main groups:

1 Tubercle bacilli: *Mycobacterium tuberculosis* (Plate 27) and *M. bovis*

2 Leprosy bacillus: *M. leprae*

3 *Atypical mycobacteria*: some tuberculosis-like illnesses in humans are caused by other species of mycobacteria. They are sometimes also referred to as mycobacteria other than tuberculosis (MOTT). They can grow at 27°C, 42°C or 45°C; some produce pigment when growing in light and are called photochromogens, whereas others produce pigment in light or darkness and are referred to as scotochromogens. Unlike *M. tuberculosis*, other mycobacteria can be rapid growers. All these species of mycobacteria are commonly referred to as the atypical mycobacteria; examples include *M. kansasii* (photochromogenic), *M. avium-intracellulare* (non-pigmented) and *M. chelonei*, (fast growing).

Cell-wall-deficient bacteria

Some bacteria do not form cell walls and are called mycoplasmas. Pathogenic species include *Mycoplasma pneumoniae* and *Ureaplasma urealyticum*. It is important to distinguish mycoplasmas from other cell-wall-deficient forms of bacteria, which can be defined as either L-forms or protoplasts:

• *L-forms* are cell-wall-deficient forms of bacteria, which are produced by removal of a bacterium's cell wall, e.g. with cell-wall-acting antibiotics such as the β-lactams. L-Forms are able to multiply and their colonial morphology is similar to the 'fried egg' appearance of the mycoplasmas.

• *Protoplasts* are bacteria that have also had their cell walls removed. They are metabolically active and can grow, but are unable to multiply. They survive only in an osmotically stabilized medium.

Chapter 5

Staphylococci

There are at least 20 species of staphylococci that are associated with humans; however, only three are clinically important: *Staphylococcus aureus, S. epidermidis* and *S. saprophyticus*. Their principal characteristics are shown in Table 5.1.

Definition
Gram-positive cocci; usually arranged in clusters (Plate 3); non-motile; catalase positive; non-sporing; grow over a wide temperature range (10–42°C) with an optimum of 37°C; aerobic and facultatively anaerobic; grow on simple media.

Classification
1 *Colonial morphology*: *S. aureus* colonies are grey to golden yellow (Plate 4); *S. epidermidis* and *S. saprophyticus* colonies are white. *S. aureus* may produce haemolysins, resulting in haemolysis on blood agar.
2 *Coagulase test*: *S. aureus* possesses the enzyme coagulase, which acts on plasma to form a clot. Other staphylococci (e.g. *S. epidermidis* and *S. saprophiticus*) do not possess this enzyme and are often termed, collectively, 'coagulase-negative staphylococci'. There are two methods to demonstrate the presence of coagulase (Plate 6):
   (a) tube coagulase test: diluted plasma is mixed with a suspension of the bacteria; after incubation, clot formation indicates *S. aureus*
   (b) slide coagulase test: a more rapid and simple method in which a drop of plasma is added to a suspension of staphylococci on a glass slide; visible clumping indicates the presence of coagulase.
3 *Deoxyribonuclease (DNAase) production*: *S. aureus* possesses an enzyme, DNAase, which depolymerizes and hydrolyses DNA; other staphylococci rarely possess this enzyme (Plate 5).
4 *Protein A detection*: *S. aureus* possesses a cell-wall antigen, protein A; antibodies to protein A agglutinate *S. aureus* but not other staphylococci.
5 Novobiocin sensitivity: useful for differentiating between species of coagulase-negative staphylococci; *S. saprophyticus* is novobiocin resistant and *S. epidermidis* is sensitive.

S. aureus

Epidemiology
S. aureus is a relatively common human commensal: nasal carriage occurs in 30–50% of healthy adults, faecal carriage in about 20% and skin carriage in 5–10%, particularly the axilla and perineum. *S. aureus* is spread via droplets and skin scales which contaminate clothing, bedlinen and other environmental sources.

Morphology and identification
On microscopy *S. aureus* is seen as typical Gram-positive cocci in 'grape-like' clusters. It is both coagulase and DNAase positive. Other biochemical tests can be performed for full identification.

Table 5.1 Characteristics of staphylococci.

TYPICAL CHARACTERISTICS OF STAPHYLOCOCCI			
Characteristic	*S. aureus*	*S. epidermidis*	*S. saprophyticus*
Coagulase	+	–	–
Deoxyribonuclease	+	–	–
Novobiocin	S	S	R
Colonial appearance	Golden-yellow	White	White, smooth, shiny
Body sites colonized	Nose	Skin	Periurethral
	Mucosal surfaces	Mucosal surfaces	Faeces
	Faeces		
	Skin		
Common infections	Skin (boils, impetigo, furuncles, wound infections	Infections of prosthetic devices, e.g. artificial valves, heart, intravenous catheters, CSF shunts	Urinary tract infections in sexually active young women
	Abscesses		
	Osteomyelitis		
	Septic arthritis		
	Septicaemia		
	Infective endocarditis		
	Device-related infections		

+, present; –, absent; CSF, cerebrospinal fluid, S, sensitive R, resistant

Pathogenicity

S. aureus produces disease because of its ability to adhere to cells, spread in tissues and form abscesses, produce extracellular enzymes or exotoxins (Tables 5.2 and 5.3), combat host defences and resist treatment with many antibiotics.

Adhesions

S. aureus has a wide repertoire of adhesions known as MSCRAMMs (microbial surface components recognizing adhesive matrix molecules) which mediate adherence to host cells; these include protein A, fibrinogen and fibronectin-binding and collagen-binding protein.

Exotoxins and enzymes

• *Coagulase: S. aureus* produces coagulase, an enzyme that coagulates plasma. Coagulase results in fibrin deposition which interferes with phagocytosis and increases the ability of the organism to invade tissues.

• *Other enzymes: S. aureus* may also produce staphylokinase (results in fibrinolysis), hyaluronidase (dissolves hyaluronic acid), proteases (degrade proteins) and lipases (solubilize lipids).

• *Haemolysin, leukotoxin and leukocidin:* several exotoxins are produced by *S. aureus*; α-toxin (haemolysin) lyses erythrocytes and damages platelets; β-toxin degrades sphingomyelin and is toxic for many types of cell, including erythrocytes; leukocidin (Panton Valentine leukocidin) lyses white blood cells and damages membranes and susceptible cells.

• *Enterotoxins:* there are six soluble enterotoxins that are produced by almost half all *S. aureus* strains. They are heat stable (resistant at 100°C for 30min), unaffected by gastrointestinal enzymes and are a cause of food poisoning, principally associated with vomiting.

• *Exfoliative/epidermolytic toxin:* some strains produce a toxin that can result in generalized desquamation of the skin (staphylococcal scalded skin syndrome).

• *Toxic shock syndrome toxin (TSST):* this is associ-

Table 5.2 Toxins produced by *Staphylococcus aureus.*

TOXIN PRODUCTION	
Toxin*	**Effect**
Enterotoxins	Released into foods and results in food poisoning with profuse vomiting
Exfoliative or epidermolytic	Scalded skin syndrome, producing peeling of layers of epidermis
Haemolysins	Lyse erythrocytes
Leukocidins	Lyse leucocytes and macrophages
Toxic shock syndrome (TSST)	Vascular collapse, rash with desquamation, systemic shock

* Toxin production varies between strains of *S. aureus.*

Table 5.3 Pathogenicity factors produced by *S. aureus.*

Factor	Effect
MSCRAMMs	Mediate adherence to host cells
Protein A	Evade host defence/inhibits phagocytosis
Fibronectin-binding protein	Mediates binding to fibronectin
Fibrogen-binding protein	Clumping factors
Capsule	Evade host defences
Coagulase	Generates protective fibrin layer around *S. aureus*
Staphylokinase	Fibrinolysis
Proteases	Degrade antibacterial proteins and matrix proteins
Lipases	Promote interstitial spreading of organism
Hyaluronidase	Degrades hyaluronic acid
α-Haemolysin	Lyses erythrocytes, damages platelets
β-Haemolysin	Degrades sphingomyelin/toxic for cells
Leukocidin/leucotoxin	Lyse white blood cells
Exotoxins, e.g. enterotoxins	Food poisoning with profuse vomiting
Superantigens, e.g. TSST, exfoliative toxin	Toxic shock syndrome, scalded skin syndrome

ated with shock and desquamation of skin, and is usually related to an underlying *S. aureus* infection. Staphylococcal enterotoxins, TSSTs and exfoliative toxin are 'superantigens', all of which bind non-specifically to specific white cells resulting in over production of cytokines giving rise to a toxic shock-like presentation.

Cell envelope
Over 90% of all clinical isolates of *S. aureus* strains possess a polysaccharide capsule that interferes with opsonization and phagocytosis. *S. aureus* also possesses a cell-wall protein (protein A) that binds the Fc component of the antibody, preventing complement activation.

Antibiotic resistance
Many strains of *S. aureus* are resistant to the antibiotic meticillin and are termed 'methicillin-resistant *S. aureus*' (MRSA). Most resistance depends on the production of an additional penicillin-binding protein, which is encoded by an acquired *mecA* gene. Many strains of MRSA are now resistant to multiple antibiotics.

Laboratory diagnosis
Laboratory diagnosis is by microscopic detection of the organism in clinical samples, direct isolation from the infected site or blood cultures, and detection of serum antibodies to staphylococcal haemolysin and DNAase. *S. aureus* strains can be

typed ('fingerprinted') by conventional methods including biotype and antibiogram, and also by the use of bacteriophages (phage typing). *S. aureus* can also be genotyped by molecular methods including pulsed field gel electrophoresis (PFGE). Typing of *S. aureus* is useful in epidemiological studies.

Antibiotic therapy
Antimicrobial agents such as flucloxacillin with or without fusidic acid remain the first-line treatment for sensitive strains of *S. aureus*; however, the increase in infections caused by MRSA has required the use of glycopeptide antibiotics such as vancomycin. MRSA can cause sepsis, ranging from wound infections to urinary tract infections and septicaemia. Epidemic strains of MRSA (EMRSA) have also been recognized. Prevention of spread through effective infection control procedures is therefore important.

Associated infections
- Skin: boils, impetigo, furuncles, wound infections, staphylococcal scalded skin syndrome.
- Respiratory: pneumonia, lung abscesses, exacerbations of chronic lung disease.
- Skeletal: most common cause of osteomyelitis and septic arthritis.
- Invasive: septicaemia (including infective endocarditis), deep abscesses (brain, liver, spleen), toxic shock syndrome.
- Gastrointestinal: toxin-mediated food poisoning.
- Device related: indwelling catheters, prosthetic joints

S. epidermidis

- *S. epidermidis* is both coagulase and DNAase negative and is present in large numbers on the human skin and mucous membranes.
- *S. epidermidis* is a cause of bacterial endocarditis, particularly in patients with prosthetic heart valves and in drug addicts. It is also a major cause of infections of implanted plastic devices such as cerebrospinal shunts, hip prostheses, central venous and peritoneal dialysis catheters.
- The organism colonizes plastic devices by attaching firmly on to artificial surfaces (Plate 1). Some strains also produce a slime layer (glycocalyx) which appears to facilitate adhesion and protect the organism from antibiotics and host defences. The increased use of implanted devices, particularly central venous catheters, has resulted in *S. epidermidis* becoming one of the most frequently isolated organisms from blood cultures. *S. epidermidis* occasionally causes urinary tract infections, particularly in catheterized patients. When isolated from hospitalized patients *S. epidermidis* is often resistant to antibiotics such as flucloxacillin and erythromycin, necessitating the use of glycopeptide antibiotics (e.g. vancomycin).

S. saprophyticus

S. saprophyticus is both coagulase and DNAase negative and also novobiocin resistant. *S. saprophyticus* is frequently associated with urinary tract infections in sexually active young women, occasionally resulting in severe cystitis ('honeymoon cystitis') with haematuria.

Chapter 6

Streptococci and enterococci

Streptococci are Gram-positive cocci, which are arranged in pairs or chains (Plate 7). They are commensals of the human upper respiratory, gastrointestinal and female genital tracts. Certain species are pathogenic. Table 6.1 lists the major infections caused by streptococci.

Note: *Streptococcus faecalis* (faecal streptococci; enterococci) has now been placed in a separate genus, *Enterococcus*, and is described at the end of this chapter together with the anaerobic streptococci (*Peptococcus* and *Peptostreptococcus*).

Definition
Gram-positive cocci; non-motile; non-sporing; occasionally capsulate; optimum growth at 37°C; facultative anaerobes; often require enriched media; many species show characteristic haemolysis on blood agar.

Classification
The classification of streptococci is based on the following factors.
1 *Haemolysis*: streptococci are alpha (α), beta (β) or non haemolytic.
   (a) α-Haemolysis is the formation around colonies on blood agar of a greenish zone in which erythrocytes are disrupted and the haemoglobin converted to a green pigment (Plate 8). Examples of α-haemolytic streptococci include the viridans streptococci and *S. pneumoniae*.
   (b) β-Haemolysis is the formation around

colonies on blood agar of a clear zone in which erythrocytes have been completely lysed (Plate 9). The degree of haemolysis can vary between strains. β-Haemolytic streptococci are classified further according to cell-wall antigens.
   (c) Non-haemolytic streptococci are unable to haemolyse erythrocytes and have no affect on blood agar.
2 *Cell envelope antigens*: in the 1930s, Rebecca Lancefield classified β-haemolytic streptococci alphabetically according to the possession of specific carbohydrate antigens in the cell envelope. Specific antibodies to each of the streptococcal group antigens were used in agglutination tests. This still forms the basis of modern laboratory techniques for the routine grouping of β-haemolytic streptococci. The medically important β-haemolytic streptococci groups are A, B, C, F and G. Group D streptococci (previously called *S. faecalis*) have been reclassified in the genus *Enterococcus*.
3 *Biochemical reactions*: some streptococci are difficult to classify by haemolytic and antigenic characteristics; biochemical tests are used in their full identification.

β-haemolytic streptococci

Group A β-haemolytic streptococci (*S. pyogenes*)
Epidemiology
Group A β-haemolytic streptococci are upper

Table 6.1 Some infections caused by streptococci.

INFECTIONS CAUSED BY STREPTOCOCCI	
Streptococcus	**Major clinical syndromes**
β-*Haemolytic*	
Group A	Pharyngitis
	Cellulitis, erysipelas, necrotizing fasciitis
	Septicaemia
	Rheumatic fever (immunologically mediated)
	Acute glomerulonephritis (immunologically mediated)
	Scarlet fever (toxin mediated)
Group B	Neonatal meningitis and septicaemia
Group C	Pharyngitis
	Cellulitis
Group G	Cellulitis
α-*Haemolytic*	
Viridans streptococci	Dental caries
	Endocarditis
S. pneumoniae	Otitis media/sinusitis
	Meningitis
	Pneumonia
Others	
S. milleri	Deep abscesses
Anaerobic streptococci	Abscesses

respiratory tract commensals of 3–5% of adults and up to 10% of children. Transmission is principally via droplet spread.

Morphology and identification

These streptococci are facultative anaerobic organisms that grow best on enriched media containing blood; typically they show a large zone of clear β-haemolysis around colonies. The identity is confirmed by Lancefield grouping and biochemical reactions.

Pathogenicity

Group A β-haemolytic streptococci demonstrate a range of virulence factors:

• *Fimbriae/pili:* facilitate adherence to pharyngeal epithelial cells and consist of lipoteichoic acid and M-protein.

• *Lipoteichoic acid*: an adherence factor.

• *M-proteins:* surface proteins that act as anti-phagocytic factors and also bind host proteases, facilitating spread through tissue. Different strains possess different M-proteins and this may be used for epidemiological typing. Antibodies to M-proteins are protective against infection.

• *Haemolysins and leukocidins:* exotoxins that lyse erythrocytes and leukocytes (streptolysins). Streptolysin O is cardiotoxic and also responsible for β-haemolysis seen on blood agar.

• *Exotoxins:* include streptokinase (which facilitates the breakdown of fibrin and allows spread of streptococci through tissues), hyaluronidase (which breaks down hyaluronic acid in connective tissue facilitating spread) and deoxyribonucleases.

• *Streptococcal pyrogenic exotoxin A, B, C, D, F (erythrogenic toxin):* responsible for the rash of scarlet fever. Production is dependent on the presence of a bacteriophage which codes for the toxin. Streptococcal pyrogenic exotoxins are 'superantigens' which facilitate massive release of cytokines, potentially leading to shock.

• *F-proteins:* surface protein that binds to fibronectin (component of host's extracelluar matrix).

- *C5a peptidase:* cleaves and inactivates the chemotactic component C5a of the complement system, limiting the recruitment of neutrophils to the site of infection.
- *Hyaluronic acid capsule:* molecular mimicry—avoidance of host defences; facilitates adhesions to cellular receptors.
- *Streptococcal inhibitor of complement (SIC):* inhibits human complement (C5b–C9).

Associated infections
- *Upper respiratory tract*: pharyngitis, tonsillitis and otitis media.
- *Skin*: cellulitis, necrotizing fasciitis, impetigo, erysipelas, wound infection and scarlet fever.
- *Invasive*: septicaemia and puerperal sepsis.

Post-infection complications
Rheumatic fever and acute glomerulonephritis are post-streptococcal, immune-mediated complications, which develop up to 3 weeks after the primary infection. In rheumatic fever a cross-reaction between group A streptococcal cell-wall antigen and cardiac muscle occurs. Streptolysin O (cardiotoxin) may also have a direct effect on heart tissue. In acute post-streptococcal glomerulonephritis, immune complexes of streptococcal antigen and antibody are deposited in the glomeruli; the resulting complement activation and inflammatory response cause glomerulonephritis.

Laboratory diagnosis
- Diagnosis is by isolation of the organism from infected sites (e.g. throat, skin) or blood cultures.
- Retrospective diagnosis is by detection of serum antibodies to streptolysin O (ASO test). This is important in the diagnosis of rheumatic fever and acute glomerulonephritis because the organism is often no longer present at the time of presentation.
- M-proteins of group A β-haemolytic streptococci can be used for epidemiological typing.

Antibiotic therapy
Penicillin (erythromycin for penicillin-allergic patients) remains the drug of choice for treatment of infection by group A β-haemolytic streptococci.

Group B β-haemolytic streptococci (*S. agalactiae*)

Epidemiology
Group B β-haemolytic streptococci form part of the normal perineal flora in about 30–40% of individuals.

Morphology and identification
These streptococci grow readily on blood agar and are identified by Lancefield grouping.

Pathogenicity
The virulence factors are less well defined than for group A β-haemolytic streptococci. Type-specific antigens (carbohydrate and protein) appear to be important in virulence (analogous to M-proteins of the group A streptococci).

Associated infections
- *Neonates*: group B streptococci are an important cause of septicaemia and meningitis.
- *Adults*: they are an occasional cause of urinary tract and genital infections, postpartum sepsis, septicaemia (particularly in elderly people) and endocarditis.

Laboratory diagnosis
This is by isolation from the infected site, e.g. blood, cerebrospinal fluid (CSF). Direct detection of the antigen from CSF or urine by a monoclonal antibody-agglutination reaction may be performed.

Antibiotic therapy
Penicillin is the drug of choice for treatment of group B β-haemolytic streptococcal infection.

Other β-haemolytic streptococci

Group C streptococci contain a number of different species, some of which are important animal pathogens. Infections in humans occur and are similar to those caused by group A streptococci.

Group G streptococci occasionally cause skin infections in humans.

α-haemolytic streptococci

The α-haemolytic streptococci are classified into *S. pneumoniae* (pneumococcus) and the group 'viridans' streptococci.

S. pneumoniae

Epidemiology

S. pneumoniae is a normal inhabitant of the upper respiratory tract.

Morphology and identification

Gram-positive cocci, often in pairs with a capsule. They are facultative anaerobic organisms with enhanced growth in the presence of 10% carbon dioxide (CO_2). Colonies are typically disc shaped with raised edges ('draughtsmen') (Plate 10) surrounded by a zone of α-haemolysis. Pneumococci are identified by their sensitivity to optochin.

Pathogenicity

Pneumococci possess a polysaccharide capsule of which there are over 80 antigenic types. Type-specific antibody to capsular antigen is required by the host for efficient opsonization and phagocytosis of the organism. Only a few serotypes account for the majority of the infection. Pneumococci also produce IgA1 protease, which cleaves IGA1 antibody, and the exotoxin pneumolysin, which inhibits neutrophil chemotaxis, phagocytosis and immunoglobulin synthesis.

Associated infections

• Respiratory tract: otitis media and sinusitis; *S. pneumoniae* is the most common cause of lower respiratory tract infection including pneumonia.
• It is an important cause of bacterial meningitis, particularly in elderly people.
• Patients who have had splenectomies or whose spleen is functionally impaired have a reduced capacity to produce IgG antibodies to carbohydrate antigens and are particularly susceptible to invasive pneumococcal disease.

Laboratory diagnosis

This is by microscopy and culture of specimens from the infected site, e.g. sputum, CSF (Plate 11) or blood. Direct detection of pneumococcal antigen in specimens (CSF, sputum, urine) can be made by an agglutination reaction and is particularly useful in the rapid diagnosis of pneumococcal meningitis and septicaemia. Serological tests for the detection of antibody to pneumococci are of limited use because of the many pneumococcal serotypes.

Treatment and prophylaxis

Until recently, pneumococci have remained sensitive to penicillin and erythromycin. However, penicillin-resistant strains have now emerged and are a particular problem in South Africa and parts of Europe. In some areas of the UK, up to 5% of strains are resistant to penicillin.

A vaccine containing 14 of the most commonly isolated pneumococcal serotypes can be given to patients at particular risk of associated infections, including those with sickle cell disease, asplenia, other forms of immunodeficiency, or chronic disease including renal, heart, lung or liver disease. Penicillin prophylaxis may also be given either instead of, or in addition to, immunization.

Viridans streptococci

The viridans streptococci are α-haemolytic streptococci; there are a number of different species including *S. sanguis*, *S. mutans*, *S. mitis*, *S. salivarius* and *S. bovis*.

Epidemiology

The viridans streptococci are commensals of the human upper respiratory and gastrointestinal tracts. Large numbers are found in the oral cavity.

Morphology and identification

They are facultative anaerobic organisms. Colonies show zones of α-haemolysis and, unlike *S. pneumoniae*, are resistant to optochin.

Pathogenicity

Viridans streptococci possess few virulence factors. Attachment to tooth enamel and gums via various carbohydrates is important in establishment and maintenance of colonization. The ability to produce acid, particularly by *S. mutans*, has been implicated in the development of dental caries.

Associated infections
• *Dental caries*: the factors involved in the pathogenesis of dental caries are complex; the viridans streptococci, particularly *S. mutans*, have been implicated in this process.
• *Bacterial endocarditis*: viridans streptococci are a common cause of bacterial endocarditis (now second to staphylococci); the organism may enter the bloodstream as a result of dental manipulation and attach to damaged cardiac valves.

Treatment
The viridans streptococci are usually sensitive to penicillin and erythromycin. The treatment of bacterial endocarditis caused by more resistant organisms may require antibiotic combinations, e.g. penicillin plus gentamicin.

S. milleri

S. milleri, an important cause of deep-seated abscesses, does not fall neatly into the normal classification of streptococci. It is often α-haemolytic but may be non- or β-haemolytic; strains often contain Lancefield group F antigen, but others may not be groupable or belong to other Lancefield groups. *S. milleri* is found in the normal human gut and is a common cause of abscesses in the abdominal cavity, chest and brain. It is often found in association with other bacteria.

Culture properties are similar to other streptococci. The organism is sensitive to penicillin but treatment of deep-seated abscesses often requires drainage and additional antibiotics because of the presence of other bacteria.

Enterococci

• Enterococci have similar properties to streptococci, but differ in their ability to grow on bile-salt-containing media, e.g. MacConkey's medium.
• They grow well on blood agar, displaying α- or β-haemolysis; some strains are non-haemolytic. The majority of strains belong to Lancefield group D. The important species are *E. faecalis* and *E. faecium*, which can be distinguished by biochemical tests.
• The principal habitat of enterococci is the gastrointestinal tract. Enterococci cause a number of important infections, including endocarditis, urinary tract and wound infections, and abscesses (particularly in association with coliforms and anaerobic bacteria).
• Unlike most streptococci, enterococci, particularly *E. faecium*, have developed resistance to penicillin. Furthermore, the emergence of enterococci resistant to glycopeptide antibiotics (e.g. vancomycin) is of particular concern in the clinical setting. These bacteria are termed 'vancomycin-resistant enterococci' (VRE) and some strains may be resistant to several antibiotics. Occasionally enterococci which require vancomycin to grow have been isolated (Plate 12).

Anaerobic streptococci

Strictly anaerobic streptococci include the genera *Peptococcus* and *Peptostreptococcus*. They are commensals of the bowel and vagina; and cause abscesses, normally in association with other bacteria. They are sensitive to penicillin and metronidazole.

Chapter 7

Clostridia

There are four main clostridial species of medical importance: *Clostridium perfringens*, *C. tetani*, *C. botulinum* and *C. difficile*. Table 7.1 lists infections caused by these clostridia.

Definition
Gram-positive bacilli (Plate 13 and Plate 2); strict anaerobes; spore forming; grow on simple media; many species produce exotoxins.

Epidemiology
Spore formation (Plate 2) allows survival in hostile environments. The ability to survive many forms of heat and chemical disinfection is important in sterilization procedures. Both spores and vegetative forms are found in soil and the gastrointestinal tracts of mammals.

Classification
Clostridia can be distinguished by morphology, biochemical reactions and antigenic characteristics.

C. perfringens

Morphology and identification
● Gram-positive rod with subterminal oval-shaped spores that form irregular colonies, often surrounded by a double zone of β-haemolysis on blood agar. The inner zone of complete haemolysis is caused by θ-toxin whereas the outer zone of par-

tial haemolysis is caused by α-toxin. *C. perfringens* ferments a variety of sugars with the production of gas.
● Five types of *C. perfringens* (A–E) are recognized based on surface antigens and the major types of lethal toxins produced:
  1 Type A strains, most commonly found in human infections, producing only the α-toxin.
  2 Types B–E, usually associated with disease in various animals, including lambs, sheep, goat and cattle, producing α-toxin plus additional toxins.

Pathogenicity
Related to toxin production; the α-toxin is responsible for the toxaemia associated with gas gangrene. The characteristics of α-toxin include lecithinase, haemolytic and necrotizing.

Associated infections
● Often associated with ischaemia (e.g. peripheral vascular disease; diabetes mellitus).
● Anaerobic cellulitis.
● Gas gangrene (Plate 67).
● Food poisoning.

Laboratory diagnosis
Isolation from infected tissue or blood cultures. Cases of food poisoning can be shown by isolation of the organism in consumed food and in the faeces. Identification of *C. perfringens* can be con-

Table 7.1 Infections caused by clostridia.

INFECTIONS CAUSED BY CLOSTRIDIA	
Species	**Infection**
C. perfringens	Gas gangrene, intra-abdominal sepsis
	Anaerobic cellulitis
	Food poisoning, postabortion infection, septicaemia
C. tetani	Tetanus
C. botulinum	Botulism, infant botulism, wound botulism
C. difficile	Pseudomembranous colitis, antibiotic-associated diarrhoea

firmed by biochemical reactions and the detection of α-toxin (lecithinase) and lipase production on egg yolk agar (Nagler plate).

Toxin-producing strains produce a zone of opalescence around growth on egg-yolk-containing media; this effect is inhibited by specific antitoxin to α-toxin (Plate 14).

Treatment and prevention
In the developed world, gas gangrene has largely disappeared as a result of changes in surgical practice with débridement of devitalized tissue. Penicillin prophylaxis is normally prescribed for amputation of the lower limb, particularly when the blood supply is poor, to prevent clostridial infections. Treatment of established gas gangrene involves surgical débridement, high-dose benzylpenicillin and supportive measures. Other antibiotics including metronidazole are active against *C. perfringens*.

C. tetani

Morphology and identification
Gram-positive rods with spherical terminal spores ('drum stick;, see Plate 2). They are motile and form a fine film of growth without discrete colonies on blood agar.

Pathogenicity
C. tetani produces a powerful neurotoxin, tetanospasmin, which binds to ganglioside receptors of neurons and blocks neurotransmitter release.

Associated infections
Tetanus may arise when *C. tetani* spores contaminate a wound. Spores germinate under anaerobic conditions (e.g. devitalized tissue); the organism multiplies and produces a powerful neurotoxin, resulting in spastic paralysis.

Laboratory diagnosis
Diagnosis is by direct isolation of the organism from affected tissue or by detection of circulating toxin by immunoassay.

Treatment and prevention
Although tetanus is rare in the developed world, it remains a major problem worldwide. A tetanus toxoid vaccine is available and is important in prevention. It offers protection for at least 10 years. Treatment includes human tetanus immunoglobulin, débridement of any wounds and benzylpenicillin to eradicate any foci of infection.

C. botulinum

Morphology and identification
Gram-positive rods, motile with subterminal oval spores. Several types (A–F) are recognized, each producing an antigenically distinct exotoxin.

Pathogenicity
The organism produces a neurotoxin that blocks release of acetylcholine at neuromuscular junctions, resulting in a flaccid paralysis.

Associated infections (rare in the UK)
• Human botulism is normally caused by types A,

B and D. Spores contaminate food (particularly preserved foods) and germinates. The organism then replicates and produces toxin. Ingestion of the toxin results in a flaccid paralysis. It may also be caused by wound contamination and toxin production (cases reported in intravenous drug abusers).

- Infant botulism is a rare cause of flaccid paralysis in infants and is associated with overgrowth of *C. botulinum* in the gut.

Laboratory diagnosis
By detection of toxin in blood, faeces or food.

Treatment and prevention
Respiratory support and antitoxin; wound and infant botulism treated with penicillin or metronidazole to stop toxin production. Prevention is dependent on good practice in food preparation and preservation.

C. difficile

C. difficile causes antibiotic-associated diarrhoea (AAD) and pseudomembranous colitis (PMC), a complication of antibiotic therapy.

Morphology and identification
On microscopy *C. difficile* appears as a large Gram-positive bacillus with terminal spores. It produces characteristic irregular shaped colonies on blood agar (Plate 15).

Epidemiology
Present in the colon of about half of all healthy neonates during the first year of life and about 20% of healthy adults; antibiotics including ampicillin, clindamycin and various cephalosporins may suppress normal bowel flora, allowing overgrowth of *C. difficile* and subsequent toxin production. Person-to-person spread may occur in hospitals, resulting in outbreaks. *C. difficile* is the most frequently identified cause of hospital-acquired diarrhoea.

Pathogenicity
Produces at least three potential virulence factors: (1) toxin A (enterotoxin) acts on gut mucosa; (2) toxin B (cytotoxin) cytopathic effect on tissue culture cells; and (3) substance that inhibits bowel motility. A hypervirulent strain of *C. difficile* has also been identified (ribotype 027). This strain produces toxins at higher levels than normal and is also resistant to commonly used quinolone antibiotics.

Laboratory diagnosis
- The organism can be isolated from the faeces by anaerobic culture on selective media containing cycloserine cefoxitin and fructose (CCFA). Toxin production is confirmed by exposure of tissue culture cells to culture filtrates of the organism.
- Diagnosis can also be made by direct detection of toxin in faeces by:
    (a) incubation of tissue culture cells with faecal filtrates; typical cytopathic effect can be recognized within 48 h; specificity is confirmed by absence of cytopathic changes in the presence of specific antitoxin
    (b) commercial assays (enzyme-linked immunosorbent assay or ELISA (Plate 52), latex agglutination), based on specific antitoxin available.

Treatment
Oral vancomycin or intravenous or oral metronidazole.

Chapter 8

Other Gram-positive bacteria

Corynebacteria

The genus *Corynebacterium* currently consists of many species most of which are human commensals. The main human pathogens comprise *Corynebacterium diphtheriae*, *C. ulcerans* and *C. jeikeium*.

Definition

Gram-positive, club-shaped rods; non-capsulate; non-sporing; aerobic; growth at optimum of 37°C on basic media; growth improved by addition of serum or blood.

C. diphtheriae

C. diphtheriae is the cause of diphtheria, an upper respiratory tract infection with complicating cardiac and neurological pathology. It is rare in the UK (< 5 cases per year).

Morphology and identification

- *C. diphtheriae* is a Gram-positive rod. Incomplete separation of cell walls during division leads to typical 'Chinese letter' morphology.
- When isolating *C. diphtheriae* from throat swabs, a medium containing potassium tellurite (e.g. Hoyle's tellurite) is often used; tellurite suppresses the growth of many upper respiratory tract commensals, but has no effect on the growth of *C. diphtheriae* which produces typical grey–black colonies

(Plate 16). An alternative medium for culturing *C. diphtheriae* is Tinsdale medium with added horse serum.

- *C. diphtheriae* comprises four biotypes: *var. gravis*, *var. mitis*, *var. intermedius* and *var. belfanti*. These biotypes can be distinguished by biochemical tests, haemolytic activity and antigenic characteristics.

Epidemiology

Sources of infection are from infected individuals or carriers of toxigenic strains of *C. diphtheriae*. Both the nose and throat are sites of carriage and transmission is by droplet spread.

Pathogenicity

- Toxigenic strains of *C. diphtheriae* are not invasive, but multiply locally in the pharynx and produce an exotoxin that destroys epithelial cells, resulting in an acute inflammatory reaction with a membrane formation. The main clinical manifestation is an upper respiratory tract illness with sore throat, thick membrane (Plate 58), bullneck, low-grade fever, dysphagia and headache. Toxin can cause chronic cutaneous ulceration. It may also be absorbed into the bloodstream, causing damage to myocardial, neural and kidney cells.
- A gene (*tox* gene) coding for toxin production is carried by a bacteriophage and is integrated into the bacterial chromosome.
- As with many bacterial exotoxins, diphtheria toxin has two components: B fragment,

responsible for binding to the target cell, and A fragment, responsible for target cell death by inhibition of protein synthesis.

• All biotypes with the exception of *var. belfanti* may produce the lethal diphtheria exotoxin.

Laboratory diagnosis
• Diagnosis is by direct isolation of the organism from throat swabs, preferably on tellurite medium (Plate 16).
• Toxigenicity testing: toxigenic strains can be confirmed by the conventional Elek test, which is a precipitation reaction in agar with specific antibody to diphtheria toxin. Other methods for toxigenicity testing of *C. diphtheriae* include enzyme immunoassay (EIA) and polymerase chain reaction (PCR)-based techniques.
• Identification: presumptive identification of suspected isolates of *C. diphtheriae* includes the biochemical tests catalase, urease, nitrate, pyranzinamidase and cystinase. The most commonly used commercial identification system is the API (RAPID) Coryne system (BioMerieux).
• As a result of the rarity of diphtheria in the developed world, many laboratories no longer screen routinely for the presence of *C. diphtheriae* in throat swabs. It is important to inform the laboratory if a clinical diagnosis of diphtheria is suspected.

Treatment and prevention
Treatment is with diphtheria antitoxin, and penicillin or erythromycin to eradicate the organisms. Contacts should be given antibiotic prophylaxis (erythromycin) and a full immunization course or booster dose of vaccine, depending on immunization history. Patients need to be isolated. Diphtheria immunization is part of the routine immunization of children in most developed countries. Travellers to areas of the world where diphtheria is still common should receive a vaccine booster dose.

Other corynebacteria

C. ulcerans, is a close relative of *C. diphtheriae* but it can be distinguished biochemically. It is a rare cause of pharyngitis and skin ulcers. Strains of *C. ulcerans* producing diphtheria toxin have been isolated, which cause a clinical syndrome identical to that caused by toxin-producing strains of *C. diphtheriae*. However, this is rare.

A number of other members of the genus *Corynebacterium* occasionally cause infections. These are part of the skin flora of humans, and are generally referred to as 'diphtheroid bacilli' or 'diphtheroids'. They can cause infections of indwelling intravascular catheters and other prostheses; one particular species, *C. jeikeium*, recognized by its resistance to many antibiotics, appears to be associated particularly with such infections and also endocarditis. Other corynebacteria associated with infection include: *C. pseudotuberculosis* (lymphadenitis), *C. striatum* (wound and device infections) and *C. urealyticum* (urinary tract infections).

Bacillus

The genus *Bacillus* contains two important human pathogens: *B. anthracis*, the causative organism of anthrax, and *B. cereus*, which is associated with food poisoning.

Definition
Gram-positive bacilli; spore forming; some species capsulate; aerobic; grow over wide temperature range on simple media.

B. anthracis

Epidemiology
B. anthracis causes infections principally in animals and is endemic in some areas of the world. Human infections are traditionally classified as (1) non-industrial, resulting from contact with infected animals or (2) industrial, resulting from workers processing wool, bones, hides or animal products. Spores can survive in soil for long periods of time.

Morphology and identification
B. anthracis is a Gram-positive, aerobic, capsulate, non-motile bacillus ($4\,\mu m \times 1\,\mu m$) which produces

oval spores. Identification is confirmed by a variety of morphological and biochemical tests.

Pathogenicity

B. anthracis possesses an anti-phagocytic capsule and produces a powerful exotoxin. The exotoxin is composed of three parts: (1) protective antigen that forms channels in cell membranes, (2) oedema factor (adenylyl cyclase) that aids in impairment of host defences and (3) lethal factor (protease) that is associated with lysis of macrophages.

Associated infections

Infection by *B. anthracis* is rare in the UK. It is associated with handling imported animal products. Anthrax has been used in bioterrorism. Types of infection include:

● Cutaneous anthrax: an ulcerating skin lesion with necrotic centre (malignant pustule). Accounts for 99% of human anthrax infections.
● Pulmonary (inhalation) anthrax: spores are inhaled resulting in pulmonary oedema and haemorrhage followed by septicaemia and frequently death.
● Septicaemia: a complication of cutaneous anthrax or pulmonary anthrax.
● Intestinal anthrax: after consumption of contaminated meat; symptoms include abdominal pains, diarrhoea and vomiting, and bloody faeces.

Laboratory diagnosis

This is by direct isolation of the organism from skin lesions or sputum. If aerosols are likely to be generated, the work should be performed under a safety cabinet.

Treatment and prevention

Treatment is by administration of penicillin; more recently ciprofloxacin or doxycycline have been recommended. Prevention is achieved by immunization of animals and humans at high risk and by slaughter and cremation of infected animals. Immunization and antibiotic prophylaxis with ciprofloxacin or doxycycline are given to exposed people if not immunized.

B. cereus

B. cereus is an important cause of food poisoning and produces spores, which can survive cooking. The organism is particularly associated with rice dishes, pasta, cakes and sauces.

Pathogenicity is related to the production of two enterotoxins, A and B. Food poisoning by *B. cereus* can present as either a rapid-onset illness with vomiting (1–5 hours after eating contaminated food), or a diarrhoeal illness that is characterized by abdominal pains 8–16 hours after ingestion of contaminated food. It can rarely cause systemic infections including pneumonia, bacteraemia and endocarditis.

Listeria monocytogenes

Listeria monocytogenes is an important cause of meningitis, encephalitis and/or septicaemia in neonates, and elderly and immunocompromised patients (e.g. transplant recipients, or those with lymphoma or AIDS). In pregnant women, *L. monocytogenes* often causes an influenza-like bacteraemic illness which may lead to placentitis and/or amnionitis and infection of the fetus.

Definition

Gram-positive bacilli; motile; non-sporing; non-capsulate; aerobic and facultatively anaerobic; optimum growth at 37°C but will grow slowly at 4–8°C.

Epidemiology

L. monocytogenes is widely distributed in the environment in soil and water and forms part of the gastrointestinal flora of some animals, birds, fish, crustaceans and, occasionally, humans. It can contaminate a variety of food products (particularly soft cheeses), which may result in outbreaks of food poisoning.

Morphology and identification

L. monocytogenes is β-haemolytic on blood agar and can be distinguished from other Gram-positive bacilli by biochemical tests, including catalase production and aesculin hydrolysis, and by its characteristic 'tumbling' motility at room temperature in broth cultures.

Pathogenicity
L. monocytogenes is an intracellular pathogen that can survive within phagocytic cells. Invasive infections may be associated with granuloma formation.

Associated infections
• Congenital/neonatal infection: vertical infection during pregnancy (early onset) may lead to abortion or pre-term delivery. *L. monocytogenes* is an important cause of late-onset neonatal meningitis (2 weeks postpartum) and septicaemia.
• Septicaemia and meningitis may also occur in immunocompromised and, very occasionally, in immunocompetent patients.

Laboratory diagnosis
By direct isolation of the organism from blood cultures or cerebrospinal fluid; identification is based on haemolysis, tumbling motility and biochemical tests, e.g. API Listeria (BioMerieux) (Plate 49).

Treatment
Ampicillin.

Erysipelothrix

This genus contains only one species that is pathogenic to humans: *Erysipelothrix rhusiopathiae*. This is a thin, non sporulating, Gram-positive rod, which can be distinguished from *L. monocytogenes* and other Gram-positive bacilli on biochemical characteristics. *E. rhusiopathiae* is associated with a wide range of animals and causes a number of important veterinary infections. Infection in humans is a zoonosis and the organism causes cellulitis (sometimes termed 'erysipeloid'), particularly in fish handlers, butchers and others involved in handling raw meat and fish. Erysipeloid presents as a painful lesion with induration and inflammation. Infection may also present as arthritis, septicaemia and endocarditis. Penicillin is the treatment of choice. Prevention is by protective clothing.

Non-spore-forming anaerobic Gram-positive bacilli

The non-spore-forming anaerobic Gram-positive bacilli include the genera *Actinomyces* and *Propionibacterium*.

Actinomyces

A. israelii is the most common pathogenic species and causes actinomycosis.
Epidemiology and associated infections
The organism is found as a normal commensal in the mouth, colon and vagina. Presentation includes sinus tract formation, relapsing or chronic infection. Infections occur in both immunocompromised and normal hosts, including:
• abscesses in the oral–facial region often associated with recent trauma or dental extraction
• abdominal infection including abscess, many after appendicitis
• uterine infections associated with the use of intrauterine contraceptive devices
• invasive infections in the immunocompromised patient
• chest infection.

Morphology and identification
A. israelii is a Gram-positive bacillus that forms branching filaments. Filaments may aggregate to form visible granules (sulphur granules) in pus. Anaerobic or microaerophilic atmosphere is required for growth, which takes up to 10 days before colonies appear. They grow on simple media but growth is enhanced on blood agar.

Treatment
Penicillin for up to 12 months.

Propionibacteria

Propionibacteria are Gram-positive diphtheroid organisms that may bifurcate or branch (Plate 17). They constitute part of the normal flora of skin and may also reside in the intestine, mouth and on the conjunctiva. Usually considered to be non-pathogenic, they can rarely cause a variety of infections.

Predisposing factors for infections caused by propionibacteria include immunosuppression, the presence of foreign bodies, preceding surgery, trauma and diabetes.

Propionibacterium acnes is the common pathogenic species and its association with acne vulgaris is widely accepted. *P. acnes* has been associated with prosthetic joint infection, endocarditis, endophthalmitis and sarcoidosis. Infections caused by *P. acnes* are considered easy to treat because the organism is invariably susceptible to many antibiotics with the exception of metronidazole.

Non-spore-forming aerobic Gram-positive bacilli

The non-spore-forming aerobic Gram-positive bacilli include the genera *Nocardia* and *Rhodococcus*.

Nocardia

Nocardia species have features in common with *Actinomyces*; they form Gram-positive branching filaments, but grow only aerobically. *Nocardia* species are found widely in the environment and infection in humans is rare. At least 13 species are responsible for human infection. The two important species are: *N. asteroids*, which is an opportunist pathogen causing abscesses in the chest, brain, liver and other organs, particularly in immunocompromised patients; and *N. braziliensis* which causes similar infections, but mainly in patients in tropical regions of America and Australia. Transmission is usually via the airborne route.

The organism can be grown from various clinical samples.

Treatment is often difficult and requires the use of long-term antibiotics, such as sulphonamides.

Rhodococcus

The genus *Rhodococcus* contains several species, some of which may cause disease.

R. equi is the major species of medical importance. It is widely distributed in the environment and has been recovered from soil and the faeces of grazing herbivores. *R. equi* is a rare opportunistic pathogen found in severely immunocompromised patients, e.g. HIV-infected individuals. Most patients present with a slow progressive granulomatous pneumonia, which may progress to cavitating lesions. Other infections include brain and subcutaneous abscesses, and infections at other sites may occur. Infections caused by *R. equi* are difficult to treat.

Treatment involves combination of two to three antimicrobial agents. Sinus infections are treated with vancomycin or a carbapenem, rifampicin and fluoroquinolones for up to 6 months.

Tropheryma whippeli

This is the causative agent of Whipple's disease, which is a rare multisystem chronic infection primarily involving the gastrointestinal tract; symptoms include diarrhoea, malabsorption, weight loss and arthropathy. Culture of the organism is difficult. Histology can assist in establishing the diagnosis.

Treatment remains empirical: trimethoprim–sulphamethoxazole has been recommended.

Chapter 9

Gram-negative cocci

Neisseria

The genus *Neisseria* consists of Gram-negative cocci and includes two important human pathogens: *N. gonorrhoeae* (gonococcus) and *N. meningitidis* (meningococcus) (Table 9.1). Some *Neisseria* species are normal commensals of the human upper respiratory tract.

Table 9.1 Gram-negative cocci and associated infections.

GRAM-NEGATIVE COCCI	
N. meningitidis	Meningitis
	Septicaemia
N. gonorrhoea	Urethritis, cervicitis
	Epididymitis
	Pelvic inflammatory disease
	Neonatal ophthalmia
	Septicaemia
	Arthritis
Moraxella catarrhalis	Upper and lower respiratory tract infections

N. gonorrhoeae

Definition

Gram-negative kidney-shaped cocci, usually in pairs (Plate 18); aerobic; optimal growth at 37°C on complex media plus 5% CO_2.

Epidemiology

Obligate human parasite. Many women are asymptomatic and may act as a reservoir of infection; highly transmissible by sexual contact. Transmission to neonates may occur during passage through an infected birth canal. Occurs worldwide.

Morphology and identification

Identification can be made by microscopy when kidney-shaped Gram-negative diplococci are seen in pus cells. They are oxidase positive and have strict nutritional requirements for growth when cultured. In contrast, many commensal neisseriae will grow on simple media. Identification is based on colonial morphology, oxidase reaction, β-lactamase production and biochemical reactions, including carbohydrate fermentation and growth requirements. The Gonochek test to detect prolylaminopeptidase, an enzyme produced almost exclusively by gonococci, should also be performed.

Pathogenicity

Gonococci have cell-surface fimbriae (pili), which aid adherence to mucosal surfaces of the urogenital sites (cervix, urethra, rectum) and the oro- and nasopharynx, initiating infection. Other virulence factors include IgA proteases. In 1–2% of individuals, gonococci invade the bloodstream, resulting in disseminated gonococcal infection (DGI);

large joint arthritis and skin rash can develop. Gonococci can survive intracellularly within polymorphonucleocytes (PMNs) and some strains are also able to resist serum lysis. Re-infections occur because protective immunity does not develop.

Associated infections
• Genital: vaginal discharge, urethritis, cervicitis; complications include epididymitis, prostatitis, urethral stricture, pelvic inflammatory disease and sterility.
• Septicaemia and arthritis (rare) after DGI.
• Rectal: proctitis, perianal pain, discharge (asymptomatic in many).
• Eye: conjunctivitis, periorbital cellulitis, erythema.

Laboratory diagnosis
Specimens should be rapidly transported to the laboratory because gonococci die readily on drying. Diagnosis is by microscopy and culture of pus and secretions from various sites (depending upon the infection): cervix, urethra, rectum, conjunctiva, throat or synovial fluid. Gram stain of urethral and cervical discharge smears show Gram-negative diplococci (Plate 18). Clinical samples are cultured on enriched selective media and identified as above.

Blood cultures aid in diagnosing disseminated infection; serology is rarely used because of poor specificity and sensitivity. Phenotypic (auxotyping, serotyping) and genotypic (PFGE, AFLP) typing may be undertaken to differentiate between strains of gonococci.

Treatment and prevention
Penicillins: about 5% of strains produce penicillinase, but are susceptible to cephalosporins (e.g. ceftriaxone) or ciprofloxacin. Single-dose therapies are recommended to increase patient compliance. No vaccine is available. Prevention of gonorrhoea includes sex education, promotion of public awareness and the use of condoms. Contact tracing is essential in preventing further spread and the disease.

N. meningitidis

Definition
Gram-negative kidney-shaped cocci, usually in pairs; aerobic; optimum growth at 37°C on complex media in air with 5% CO_2.

Epidemiology
• *N. meningitidis* is an obligate human parasite causing meningitis and septicaemia; humans are the only natural host. The organism is carried in the upper respiratory tract by about 10% of the population. Meningococci are spread from human to human via droplets or direct contact. Sporadic cases or epidemics occur worldwide.
• At least 13 serogroups of meningococci, based on capsular polysaccharides, have been identified; the most important groups are A, B, C, D, X, Y and W-135. (In the UK, two-thirds are caused by group B and one-third by group C).
• Patients with defects of the later components of the complement system (C8, C9) may suffer recurrent meningococcal infections.
• In temperate climates most infections occur in patients aged less than 5 years or in 15–19 year olds.

Morphology and identification
• Identification is by microscopy when kidney-shaped, Gram-negative cocci are seen within polymorphonucleocytes. *N. meningitidis* is oxidase positive. Biochemical reactions (carbohydrate fermentation reactions) and Gonochek are used for identification (Plate 48).
• Serotyping is performed by reference laboratories to provide epidemiological information.

Pathogenicity
A polysaccharide capsule protects against phagocytosis and promotes intracellular survival. Cell-wall endotoxins (lipopolysaccharide) are important in the pathogenesis of severe meningococcal disease. Colonization of nasopharynx occurs; local invasion may follow with bacteraemia and meningeal involvement.

Associated infections
- Meningitis, often with septicaemia; may have purpuric skin rash.
- Septicaemia without meningitis (less common).
- Disseminated infection including; osteomyelitis, arthritis, cellulitis and pericarditis.
- Urethritis, cervicitis, proctitis after orogenital, anogenital and oroanal sexual practices.

Laboratory diagnosis
- Specimens of blood and cerebrospinal fluid (CSF) for microscopy, culture and antibiotic sensitivity; nasopharyngeal swabs are useful to determine carriage of meningococci.
- Direct detection of meningococcal antigen in CSF, serum or urine by agglutination reaction.
- Serology: antibodies to meningococcal polysaccharide antigens are assayed occasionally to provide a retrospective diagnosis.
- Polmerase chain reaction (PCR) detection of meningococci in blood and CSF.

Treatment
Penicillin or cefotaxime in penicillin-allergic patients; rifampicin or ciprofloxacin should also be given to eradicate nasopharyngeal carriage. Prevention includes antibiotic prophylaxis for close contacts. Meningococcal group C vaccine (group B vaccine being developed) and tetravalent (A, C, Y, W135) polysaccharide vaccine available for travellers to high incidence areas.

Moraxella

Moraxella catarrhalis

Definition
Gram-negative diplococci and tetrads; aerobic; optimum growth at 37°C on complex media in air with 5% CO_2.

Epidemiology
M. catarrhalis is a member of the normal flora of the upper respiratory tract. Associated with upper and lower respiratory tract infection. Transmission is a result of direct contact with contaminated secretions by droplets.

Morphology and identification
- Identification is by microscopy when numerous Gram-negative diplococci are observed.
- Colonies of *M. catarrhalis* may be characteristically nudged across the surface of the agar plates with a bacteriological loop much like a 'hockey puck'.
- *M. catarrhalis* reduces nitrates and are oxidase, catalase and DNAse positive.
- *M. catarrhalis* may be distinguished from Neisseria species by butyrate esterase detection (Tributyrin test).

Pathogenicity
Some strains of *M. catarrhalis* possess fimbriae that aid adherence to respiratory epithelium. Cell-wall endotoxin (lipopolysaccharide) may play a role in the disease process.

Associated infections
- Otitis media.
- Sinusitis
- Lower respiratory tract infections (bronchitis, pneumonia), particularly in patients with predisposing factors, including chronic obstructive pulmonary disease (COPD) or immunosuppression.

Laboratory diagnosis
- Specimens of sputum, tympanocentesis fluid and sinus aspirates for culture and microscopy.
- Serological tests are not widely used.

Treatment
Most clinical isolates produce β-lactamase. Infection may be treated with co-amoxiclav or extended broad-spectrum cephalosporins, e.g. cefotaxime.

Other Gram-negative cocci

Other Gram-negative cocci such as *N. subflava*, *N. lactamica* and *Veillonella* species, are part of the normal flora of the upper respiratory tract, and rarely cause disease.

Chapter 10

Enterobacteriaceae

The Enterobacteriaceae are a heterogeneous group of Gram-negative aerobic bacilli, which are commensals of the intestinal tract of mammals. They are also referred to as coliforms or enteric bacteria. The family includes the following important genera: *Escherichia, Enterobacter, Klebsiella, Proteus, Serratia, Shigella, Salmonella* and *Yersinia* (Table 10.1).

Definition

Gram-negative bacilli (Plate 19); aerobic and facultatively anaerobic growth; optimal growth normally at 37°C (Plate 20); grow readily on simple media; ferment wide range of carbohydrates; oxidase negative; some are motile; bile tolerant and grow readily on bile-salt-containing media, e.g. MacConkey's agar.

Morphology and identification

- Fermentation of lactose to produce pink colonies on MacConkey's agar is a property of the genera *Escherichia, Enterobacter* and *Klebsiella*. The genera *Salmonella, Shigella, Serratia, Proteus* and *Yersinia* do

Table 10.1 Enterobacteriaceae infections.

ENTEROBACTERIACEAE	
Genus/species	**Common infections**
Escherichia coli	Urinary tract infection
	Gastrointestinal infections
	Wound infections
Klebsiella	Urinary tract infection
Enterobacter	Pneumonia (aspiration)
Serratia	Wound infection
Proteus	Urinary tract infection
Salmonella serotypes *Typhi*	Enteric fever
and *Paratyphi*	Septicaemia
Other salmonellae	Enteritis
	Septicaemia
Shigella	Enteritis
Yersinia	*Y. enterocolitica*, enteritis
	Y. pestis, plague
	Y. pseudotuberculosis, mesenteric adenitis

not ferment lactose and form pale colonies on MacConkey's agar.

- Various tests are used for identification:
  (a) oxidase (−)
  (b) carbohydrate fermentation reactions
  (c) production of urease, which splits urea with release of ammonia
  (d) hydrogen sulphide production; amino acid decarboxylation and indole production.
  Commercially available kits are available for identification of the Enterobacteriaceae.
- Enterobacteriaceae possess a variety of antigens: these may include somatic or cell-wall ('O'), flagellar ('H') and capsular ('K') antigens. These are used to subdivide some genera and species, e.g. *Escherichia, Salmonella*.

Escherichia

The genus *Escherichia* currently contains five species. However, *E. coli* is the species most frequently isolated from humans.

Epidemiology and associated infections
Although *E. coli* is a commensal of the human intestine, it can cause a variety of important infections, including infections of the gastrointestinal tract, urinary tract, biliary tract, lower respiratory tract, septicaemia, haemolytic–uraemic syndrome (HUS), haemorrhagic colitis and neonatal meningitis.

Pathogenicity
- Fimbriae facilitate adherence to mucosal surfaces of the intestinal and urinary tracts.
- As with all Gram-negative bacteria, *E. coli* have lipopolysaccharides (endotoxin) in their cell walls. Endotoxin is liberated when Gram-negative bacteria lyse, resulting in complement activation, intravascular coagulopathy and endotoxic shock.
- *E. coli* has capsular antigens; the K1 antigen is associated with neonatal meningitis.
VTEC (verocytotoxin-producing *E. coli*)
- Important cause of diarrhoea and HUS.
- VTEC 0157 strain most commonly implicated.
- Most sporadic cases:
  — a zoonotic infection mainly from cattle

  — low infecting dose
  — acquired by eating undercooked contaminated food
  — damages gut endothelium, resulting in haemorrhagic colitis.
  — HUS occurs in about 5% of patients: renal failure, oliguria, thrombocytopenia.
Diarrhoea caused by other *E. coli* mainly in developing countries
- Enteropathogenic (EPEC)—cause of infantile diarrhoea.
- Enterotoxigenic (ETEC)—as EPEC, also travellers' diarrhoea.
- Enteroinvasive (EIEC)—causes dysentery-like illness.
- Enteroaggregative (EAGEC).

Laboratory diagnosis
- Diagnosis is by direct isolation of the organism from clinical samples, e.g. faeces, urine and blood.
- Identification of some pathogenic strains, e.g. VTEC, EPEC, may be achieved by serotyping.

Antibacterial therapy
E. coli is commonly resistant to penicillin and ampicillin (by production of β-lactamases). Antibiotics often used to treat *E. coli* infections include the cephalosporins, trimethoprim, ciprofloxacin and aminoglycosides; strains isolated from hospitalized patients are often more resistant to antibiotics and therefore local sensitivity patterns need to be considered.

Klebsiella

The genus *Klebsiella* contains a number of species, including *K. pneumoniae, K. aerogenes* and *K. oxytoca*; these are separated on the basis of biochemical tests. Pathogenicity is associated with capsule production and endotoxins. There are many capsular serotypes.

Klebsiellae are widespread in the environment and in the intestinal flora of humans and other mammals. Infections are often opportunist and associated with hospitalization, particularly in high-dependency units. They include pneumonia, urinary tract and wound infections, and neonatal meningitis. Outbreaks of nosocomial (hospital-acquired) infections occur.

Klebsiellae often produce β-lactamases and are resistant to ampicillin. Cephalosporins (e.g. cefotaxime and ceftriaxone), β-lactamase-stable penicillins and aminoglycosides (e.g. gentamicin and amikacin) are commonly used to treat klebsiella infections, but multiply resistant strains may limit antibiotic choice. Extensive use of broad-spectrum antibiotics in hospitalized patients has led to development of multi-drug-resistant strains that produce extended spectrum β-lactamases (ESBLs). ESBL-producing strains often possess capsular type K55 and have a propensity to spread within the clinical setting.

Enterobacter and Serratia

The genera *Enterobacter* and *Serratia* are closely related to klebsiellae. Infections occur principally in hospitalized patients and include the lower respiratory and urinary tracts. Hospital cross-infection with antibiotic-resistant strains is a particular problem. ESBL production in strains of *Enterobacter* has recently been reported.

Salmonella

The genus *Salmonella* contains a large number of species (more correctly, serotypes—see below). *Salmonella* serotype *Typhi* and *Salmonella* serotype *Paratyphi* cause enteric fever (typhoid or paratyphoid); other salmonellae cause enteritis.

Classification
• Classification of the genus *Salmonella* is complex, with about 2000 serotypes being distinguished. Many of these have been given binomial names, e.g. *Salmonella typhimurium* and *Salmonella enteritidis*, although they are not separate species, merely different serotypes. In clinical practice, laboratories identify organisms according to their binomial name.
• Salmonellae have both H (flagellar) and O (somatic) antigens. There are over 60 different O antigens, and individual strains may possess several O and H antigens; the latter can exist in different forms, termed 'phases'. *Salmonella* serotype *Typhi* also has a capsular-associated antigen referred to as 'Vi' (for virulence), which is related to invasiveness.
• Agglutination tests with antisera for different O and H antigens form the basis for the serological classification of the salmonellae. Further strain differentiation of salmonellae for epidemiological purposes can be obtained by phage typing.

Epidemiology
Salmonellae are commensals of many animals including poultry, domestic pets, birds and humans. Transmission is via the faecal–oral route. The infective dose is relatively high (approximately 10^6 organisms) and multiplication in food is important for effective transmission. A chronic carrier state can occur.

Morphology and identification
• Salmonellae are motile and produce acid, and occasionally gas, from glucose and mannose.
• They are resistant to sodium deoxycholate which inhibits many other Enterobacteriaceae, and deoxycholate agar is used as a selective media used to isolate salmonellae from stool specimens.
• Salmonellae do not ferment lactose and form pale colonies on MacConkey's medium; on xylose lysine deoxycholate (XLD) agar many salmonellae form pale colonies with black centres as a result of H_2S production. This aids recognition of salmonellae colonies in mixed cultures.
• Further biochemical tests are required for definitive identification. Serotyping of O and H antigens by slide agglutination is used for speciation.

Pathogenicity
Salmonellae can survive the acidic pH of the stomach and invade the gut, resulting in an inflammatory responses and subsequent diarrhoea. The 'Vi' antigen of *Salmonella* serotype *Typhi* is associated with invasiveness.

Associated infections
• Salmonellosis (caused by non-typhoid) salmonellae.
• Enterocolitis/gastroenteritis; rarely associated

with septacaemia, osteomyelitis, septic arthritis or abscesses.

- Enteric fever; caused by *Salmonella* serotype *Typhi* and *Salmonella* serotype *Paratyphi*. Septicaemic illness prevalent in Asia, South America and Africa; approximately 300 cases per year in the UK. May present in biliary and urinary tracts after recovery.

Laboratory diagnosis
- Enterocolitis: culture of stool samples on selective media, e.g. XLD, DCA (deoxycholate citrate agar), and enrichment media, e.g. selenite broth; identification of salmonellae by biochemical and agglutination tests. Phage typing can be used for typing individual strains.
- Enteric fever: isolation of *Salmonella* serotype *Typhi* or *Salmonella* serotype *Paratyphi* from blood cultures (first week of infection), urine (second week) or faeces (first week onwards). Serology (Widal's test) is now rarely performed because of unreliable results.

Treatment and prevention
- Enteric fever: ciprofloxacin.
- Enterocolitis: self-limiting; antibiotics (e.g. ciprofloxacin and cefotaxime) reserved for severe septicaemic cases particularly in elderly, very young and 'immunocompromised' individuals.
- Typhoid immunization; avoidance of contaminated water/food.

Shigella

Classification
The main pathogenic species are *S. sonnei*, *S. boydii*, *S. dysenteriae* and *S. flexneri*. They are distinguished by biochemical reactions and antigenic characteristics ('O' antigens).

Epidemiology
Obligate human pathogens with no animal reservoirs; transmission via faecal–oral route with low infective dose (10–200 organisms). Direct person-to-person spread is common; chronic carrier state is rare.

Morphology and identification
Shigellae are non-motile (they have no flagella). They are resistant to sodium deoxycholate and grow on deoxycholate agar (see Salmonellae). They are non-lactose or late lactose (*S. sonnei*) fermenters. Further biochemical tests are carried out for definitive identification and serotyping by slide agglutination is used for speciation.

Pathogenicity
Shigellae express an intestinal adherence factor which aids in colonization within the gut. They cause disease by invasion and destruction of the colonic mucosa, and also produce an enterotoxin (cytotoxin) known as Shiga toxin which can cause microangiopathy, HUS and thrombocytopenic purpura.

Associated infections
Many mild infections, diarrhoea, dysentery (diarrhoea with blood and pus, fever, abdominal pain) (HUS; septicaemic complications rare).

Laboratory diagnosis
Stool culture on selective media, e.g. XLD.

Treatment
Often self-limiting; antibiotics (e.g. ciprofloxacin) reserved for severe cases (often caused by *S. dysenteriae*).

Proteus

The genus *Proteus* contains a number of species, e.g. *Proteus mirabilis* and *P. vulgaris*. Characteristics include:
- non-lactose fermenting, produce pale colonies on MacConkey's agar
- motile, tendency to 'swarm' on blood agar (Plate 21)
- important cause of urinary tract and occasionally abdominal wound infections.

Yersinia

The genus *Yersinia* contains three human

pathogens: *Yersinia pestis*, *Y. pseudotuberculosis* and *Y. enterocolitica*; these species are identified by biochemical tests.

Y. pestis

Y. pestis is a cause of plague (black death). Although mainly of historical interest in Europe, plague remains endemic in some areas of the world.

It is primarily a pathogen of rodents and is transmitted to humans via infected fleas; lymph nodes associated with the flea bite enlarge to form a bubo (bubonic plague). Septicaemia and pneumonia may follow (pneumonic plague). Person-to-person spread via droplets occurs in pneumonic plague. *Y. pestis* can be isolated from blood, bubo aspiration, sputum, throat swabs and skin scrapings. Treatment for *Y. pestis* infection is with tetracycline.

Laboratory diagnosis is by microscopy and culture of clinical material.

Y. pseudotuberculosis

This is primarily an animal pathogen but in humans it is an occasional cause of mesenteric adenitis, rarely septicaemia. It is probably transmitted to humans via contaminated food.

Laboratory diagnosis is through culture of faeces on selective agar, e.g. CIN agar and cold enrichment of faeces.

Y. enterocolitica

Y. enterocolitica causes diarrhoeal disease, terminal ileitis and mesenteric adenitis. Infection may be complicated by septicaemia or reactive polyarthritis. A zoonotic infection of domestic and wild animals, transmission via faecal–oral route. Laboratory diagnosis is by culture of faeces on selective agar, e.g. CIN agar, cold enrichment of faeces, culture of blood specimens or serology.

Chapter 11

Parvobacteria

The parvobacteria are a group of genera (Table 11.1), all of which are Gram-negative coccobacilli, and often described collectively because of common characteristics. They are relatively small (0.5–1.0 μm) and grow only on enriched media. The group includes the genera *Haemophilus*, *Bordetella*, *Legionella*, *Pasteurella* and *Francisella*.

Haemophilus

Definition

Gram-negative bacilli; fastidious growth requirements; non-motile; some strains capsulate; aerobic and anaerobic growth.

H. influenzae

Classification

Strains may be capsulated or non-capsulated. Capsulated strains are divided into six serotypes (designated a–f) on the basis of polysaccharide capsular antigens; type b strains were an important cause of invasive infection, particularly in small children.

Epidemiology

Non-capsulated strains of *H. influenzae* are common commensals in the upper respiratory tract. Capsulated strains, including *H. influenzae* type b, are also carried in a small number of healthy individuals ('carriers') and are an important reservoir for invasive disease.

Morphology and identification

- *H. influenzae* is nutritionally demanding and grows only on enriched media requiring haemin (factor X) and nicotinamide adenine dinucleotide (NAD or factor V) for growth. Simple nutrient agar contains no X or V factor and strains of *H. influenzae* will grow only around paper discs containing both these factors (Plate 22).
- Growth is poor on blood agar that contains factor X but only small amounts of available factor V; heat treatment of blood before incorporation in agar produces a medium known as chocolate agar, which contains larger quantities of both factors and allows growth of *H. influenzae*.

Pathogenicity

The polysaccharide capsule of certain strains, particularly type b, is a major virulence factor. Antibodies to the type b capsule are protective for invasive *H. influenzae* type b infections. Another component of *H. influenzae* cell wall that contributes to pathogens is lipopolysaccharide (LPS) in which biological activity is similar to that of other Gram-negative endotoxins. *H. influenzae* also possesses pili (fimbriae) and expresses a surface protein called Hia, both of which facilitate adherence to cells of the respiratory tract.

Table 11.1 Common infections associated with parvobacteria.

PARVOBACTERIA		
Genus	Species	Associated infections
Haemophilus	*H. influenzae* (type b) (non-capsulated)	Epiglottitis
		Infections, including meningitis in children <5 years
		Exacerbations of chronic lung disease
	H. parainfluenzae	Exacerbations of chronic lung disease
	H. ducreyi	Genital ulcers (chancroid)
Brucella	*B. melitensis*	Brucellosis
	B. abortus	
	B. suis	
	B. canis	
Pasteurella	*P. multocida*	Cellulitis after animal bites
Legionella	*L. pneumophila*	Pneumonia, Pontiac fever
Bordetella	*B. pertussis*	Pertussis ('whooping cough')
	B. parapertussis	Pertussis-like syndrome

Associated infections

• *H. influenzae* type b strains were an important cause of serious infection in young children (aged 6 months to 5 years). Outside this age group, infections are less common.

• Infections include meningitis, epiglottitis, cellulitis, otitis media, osteomyelitis, septic arthritis and pneumonia.

• Non-type b capsulated strains occasionally cause invasive infections. Non-capsulated strains are important respiratory pathogens, particularly as a cause of acute exacerbations of chronic lung disease.

Laboratory diagnosis

• Diagnosis is by microscopic examination of clinical samples and direct isolation of the organism from infected sites, e.g. cerebrospinal fluid (CSF), sputum and blood. Identification is made by assessment of X and V dependency and biochemical tests. Serotyping by agglutination tests is carried out on capsulated strains when isolated from invasive infections.

• Direct detection of capsular antigen in CSF can be made for the rapid diagnosis of *H. influenzae* type b meningitis.

Treatment and prophylaxis

• Many strains of *H. influenzae* produce a β-lactamase and are resistant to ampicillin and many other β-lactam antibiotics. Chloramphenicol resistance is also increasing, so cefotaxime or ceftriaxone is used as first-line treatment.

• A vaccine to the type b polysaccharide capsule has led to a rapid reduction in the number of invasive *H. influenzae* type b infections. Rifampicin is used for the prophylaxis of contacts of cases of *H. influenzae* type b meningitis, and as an adjunct to therapy to reduce nasopharyngeal carriage.

H. parainfluenzae

H. parainfluenzae is non-capsulated and differs from *H. influenzae* in requiring only factor V for growth. It is a common upper respiratory tract commensal, but its role in exacerbations of chronic bronchitis and upper respiratory tract infection is ill-defined.

H. aegyptius

H. aegyptius causes an acute purulent conjunctivitis, in seasonal endemics, particularly in hot climates.

H. ducreyi

H. ducreyi causes genital ulcers (chancroid), a

sexually transmitted infection seen in Africa, Asia, Latin America and many parts of Europe.

HACEK

Some species of *Haemophilus* belong to the group of Gram-negative bacilli called HACEK (*H. parainfluenzae, H. aphrophilus, H. paraphrophilus, Actinobacillus actinomycetemocomitans, Cardiobacterium hominis, Eikenella corrodens* and *Kingella* species). This group of organisms is responsible for 5–10% of cases of infective endocarditis (IE) involving native valves. The HACEK group is also associated with infections including peritonitis, otitis media, bacteraemia and periodontal infection.

Brucella

The genus *Brucella* contains four species responsible for human infections: *Brucella melitensis, B. abortus, B. suis* and *B. canis*.

Definition
Small, Gram-negative coccobacilli; non-motile; non-capsulated; grow slowly aerobically with the addition of 10% CO_2; require enriched media including serum or blood for growth.

Epidemiology
Brucellae are intracellular parasites transmissible to a wide range of animal species including humans. Human brucellosis is therefore a zoonotic infection, the main reservoirs being goats (*B. melitensis*), cattle (*B. abortus*), pigs (*B. suis*) and dogs (*B. canis*). In the UK, *B. abortus* is the most common species isolated whereas *B. melitensis* is found most commonly in the Mediterranean area. Infection is associated with close contact with farm animals, e.g. farm workers, veterinary surgeons, or occurs as a result of ingestion of unpasteurized cows' or goats' milk. There are about 25 cases per year in the UK.

Morphology and identification
Brucella species are aerobic; they require CO_2 for growth and are slow growers. Identification is based on morphology, growth characteristics and biochemical reactions.

Pathogenicity
Brucella species are intracellular pathogens that are capable of surviving and replicating within phagocytic cells. The principal virulence factor is the cell wall LPS. Infection with *Brucella* is typical of an intracellular pathogen, with the formation of multiple granulomatous lesions in several organs.

Associated infection
Infection with *Brucella* is associated with fever and a variety of non-specific symptoms, such as malaise and weight loss. The most common sites involved are osteoarticular areas, the genitourinary tract, meninges, heart and liver.

Laboratory diagnosis
- Diagnosis is made by direct isolation from blood cultures, CSF or culture of aspirated bone marrow after prolonged incubation (2–3 weeks). *Note*: routine blood cultures are incubated normally for only 3–7 days and it is important to inform the laboratory if brucellosis is suspected.
- Serological tests including ELISA (enzyme-linked immunosorbent assay) to detect serum antibodies against *Brucella* are available; these aid diagnosis because many patients present late in their illness.

Prevention
This is by eradication of *Brucella* in the animal population, and avoidance of raw milk and associated products.

Treatment
Prolonged treatment is necessary (6 weeks) with a tetracycline and an aminoglycoside, rifampicin or co-trimoxazole. This reflects the intracellular location of the organisms.

Legionella

Legionella pneumophila and other closely related *Legionella* species are a cause of lower respiratory tract infections in humans. The organism was first recognized after investigations into an outbreak of respiratory illness in delegates at an American Legion Convention in Philadelphia, USA, in 1976.

Definition
Gram-negative bacilli; obligately aerobic, slow-growing bacteria with fastidious growth requirements; non-capsulated; motile.

Epidemiology
The organism is found in water systems, particularly heating systems and water-cooled air-conditioning plants. Although relatively fastidious in its growth requirements, the organism grows in water at temperatures between 20 and 45°C; this may be related to growth within amoebae which are found naturally in the same environment or to its ability to survive in biofilms. *Legionella* species are also relatively chlorine tolerant. Transmission occurs through aerosolization of water contaminated with legionellae. Human-to-human transmission has not been demonstrated.

Pathogenicity
Virulent *Legionella* species adhere to human respiratory epithelial cells via pili (fimbriae). Legionellae can also survive and grow within phagocytic cells, neutralizing the normal bacteria-killing mechanisms.

Associated infections
L. pneumophila can cause a range of respiratory illnesses, from atypical pneumonia to a flu-like illness with no lower respiratory symptoms, called Pontiac fever. *Legionella* may also spread to other areas of the body, including lymph nodes, brain, kidney and bone marrow. Legionellosis may be acquired both in the hospital and in the community.

Laboratory diagnosis
- Diagnosis is made by culture of sputum or bronchoalveolar lavage fluid on specialized media (BCYE); colonies take between 2 and 7 days to grow. Legionellae are classified into species according to surface antigens.
- Legionellae can be detected directly in respiratory specimens by direct immunofluorescence microscopy.
- Legionella antigen may be detected in urine samples by immunoassay.

- Serological tests are often used to confirm the diagnosis of legionellosis; a rise in serum antibody titre may take up to 6 weeks.

Treatment
This is with erythromycin combined, in severe cases, with rifampicin. Tetracyclines and fluoroquinolones are also effective.

Bordetella

The genus *Bordetella* contains four species. *B. pertussis* is the cause of whooping cough, and *B. parapertussis* has also been associated with a pertussis-like syndrome, but with less severe symptoms.

Definition
Small coccobacilli; obligate aerobes; fastidious growth requirements; non-motile; non-capsulated.

Epidemiology
B. pertussis is a human pathogen with no animal reservoir. It is readily transmissible by droplet spread from infected people. Mainly affects young children; adults can also be infected.

Pathogenicity
A variety of virulence factors has been described, including adhesins, cytotoxins and haemolysins. Lung inflammation ensues.

Associated infections
Whooping cough (pertussis) with cough and whoop; cyanosis, occasional pulmonary collapse; and haemorrhage.

Laboratory diagnosis
Culture of pernasal swabs on specialized agar, e.g. Bordet–Gengou (charcoal blood agar). Colonies may take 2–3 days to grow.

Treatment and prevention
Erythromycin may be given to reduce infectivity. If given during the incubation period, it may reduce clinical symptoms. Immunization with DTP (triple vaccine) during childhood is widely used.

Pasteurella

Pasteurella multocida is the only common human pathogen in this genus. The organism is primarily a pathogen of animals and is found as part of the normal oral flora of cats and dogs. Human infections arise after animal bites and result in a local infection with surrounding cellulitis; treatment is with penicillin. Laboratory diagnosis is by culture of the organism, which grows well on blood agar.

Chapter 12

Pseudomonas and other aerobic Gram-negative bacilli

Pseudomonas

Members of the genus *Pseudomonas* ('pseudomonads') are widely distributed in nature; *P. aeruginosa* is the most important pathogenic species in humans.

Definition
Gram-negative bacilli; strict aerobes; grow on simple media at a wide temperature range; motile; non-capsulated; some species pigmented.

P. aeruginosa

Epidemiology
- *P. aeruginosa* is an important organism in hospital-acquired infection. It is a normal commensal in the human gastrointestinal tract, but may colonize other sites when host defences are compromised, including burns and leg ulcers, the respiratory tract of patients with cystic fibrosis or bronchiectasis, and the urinary tract of patients with long-term, indwelling, urethral catheters.
- *P. aeruginosa*, as with other members of the genus, has the ability to grow with minimal nutrients, e.g. in water and in the presence of some disinfectants; these properties are the key to its important role as a hospital pathogen.

Morphology and identification
The organism is a strict aerobe that grows on most media, producing a characteristic greenish pigment. It can be distinguished from the Enterobacteriaceae by its oxidative metabolism (oxidase positive) and the inability to grow anaerobically. The different species of the genus *Pseudomonas* can be distinguished by biochemical tests.

Pathogenicity
P. aeruginosa is relatively non-pathogenic; it characteristically causes infections in hospitalized patients, particularly those who are immunocompromised. The organism produces several enzymes that allow spread through tissues (elastase) and a protease that breaks down IgA on mucosal surfaces.

Associated infections
P. aeruginosa principally causes opportunist infections including:
- skin infections associated with burns and venous ulcers
- chest infections in patients on intensive care units (ICUs) and those with cystic fibrosis
- urinary tract infections, particularly associated with long-term urethral catheterization
- septicaemia
- eye infections
- otitis externa.

Laboratory diagnosis
Isolation of the organism from relevant body sites.

The isolation of *P. aeruginosa* from several patients may suggest hospital cross-infection. In such circumstances, strains of *P. aeruginosa* can be further characterized by serotyping or molecular biological techniques.

Treatment
One of the characteristics of *P. aeruginosa* is resistance to antibiotics; some newer antibiotics have been designed specifically to combat *P. aeruginosa*. Clinically important anti-pseudomonal antibiotics include:
- aminoglycosides (e.g. gentamicin)
- broad-spectrum penicillins (e.g. piperacillin)
- third-generation cephalosporins (for example ceftazidime)
- quinolones (e.g. ciprofloxacin).

On hospital units where antibiotics are used frequently (e.g. special care baby units, ICUs), *P. aeruginosa* isolates may become resistant to these antibiotics. Isolation of patients colonized by multi-resistant *P. aeruginosa* strains is an important part of controlling hospital infections.

Burkholderia (formerly *Pseudomonas) cepacia*

This member of the genus has recently been recognized as an important cause of lower respiratory tract infection in patients with cystic fibrosis; it occurs in moist environments and hospital. Treat with meropenem.

Stenotrophomonas (formerly *Pseudomonas) maltophilia*

This organism is a cause of hospital-acquired infections and is often multiply antibiotic resistant; it causes pneumonia particularly in compromised patients, e.g. ventilated or cancer patients; also related to intravascular device infections. Treat with trimethoprim–sulfamethaxole or ticarcillin–clavulanic acid.

Acinetobacter

Acinetobacter species share a number of common features with pseudomonads. They are short, Gram-negative bacilli, strict aerobes, and grow well on simple media with minimal nutrients. However, unlike pseudomonads they are oxidase negative. They occur in many environments, including hospitals, and are an important cause of hospital-acquired infection. Hospital outbreaks, particularly chest infections, occur particularly in ICUs. *Acinetobacter* species frequently acquire plasmids expressing multiple antibiotic-resistance patterns, including quinolones, many aminoglycosides and β-lactam agents. Meropenem and amikacin may be active.

Other Gram-negative bacteria
Capnocytophaga species

A Gram-negative bacillus that grows slowly (> 5 days) on complex media and has an absolute requirement for CO_2. It is a normal commensal of the oral cavity. It causes septicaemia in patients with immunodeficiency, particularly when oral mucosal ulceration is also present, and is a rare cause of infection in immunocompetent patients (e.g. lung abscess, endocarditis).

Cardiobacterium hominis

C. hominis is a Gram-negative bacillus with tapering ends. It has a tendency to form long filaments. It grows best on enriched media in a microaerophilic environment with additional CO_2. *C. hominis* is a normal commensal of the human upper respiratory tract and is a rare cause of bacterial endocarditis.

Chromobacterium violaceum

This is a Gram-negative bacillus that grows readily on simple media to produce typical violet-pigmented colonies. The organism is saprophytic and found in soil and water, particularly in tropical and subtropical areas. It is a rare cause of pneumonia, septicaemia and abscesses.

Eikenella corrodens

E. corrodens is a Gram-negative bacillus that forms typical pitting colonies. It grows aerobically and anaerobically on complex media; colonies may take 5 days to appear. The organism is found as a normal commensal of the human upper respiratory tract and may be isolated from a variety of infections, including dental abscesses, soft-tissue infections of the neck, pulmonary and abdominal abscesses and wound infections after human bites, normally in association with other organisms such as anaerobes, coliforms and streptococci. *E. corrodens* is a rare cause of endocarditis.

Gardnerella vaginalis

G. vaginalis is a Gram-negative rod that occasionally appears Gram positive ('Gram variable'). It grows aerobically and anaerobically on complex media. It is a normal commensal of the vaginal flora, but is associated with bacterial vaginosis, a condition characterized by an 'ammonia smelling' vaginal discharge and the presence of clue cells (Plate 23) (epithelial cells with attached *G. vaginalis*).

Streptobacillus moniliformis

S. moniliformis is a short Gram-negative rod that may also appear as an L-form; it grows slowly on complex media. A commensal of healthy rats, it is one cause of rat-bite fever in humans (the other aetiological organism being *Spirillum*), an infection characterized by fever, vomiting, headache, arthralgia and a rash. *S. moniliformis* is also the cause of 'streptobacilliary fever', sometimes called Haverhill fever, after the first recorded epidemic of the infection at Haverhill, USA. Outbreaks have occurred in the UK and may be associated with the consumption of unpasteurized milk.

Chapter 13

Campylobacters, Helicobacter and Vibrios

Campylobacters

Definition

Curved, Gram-negative bacilli; motile with a polar flagellum; non-sporing; capsulated; some species show optimal growth at 42°C; require micro-aerophilic atmosphere plus 10% CO_2.

Classification

There are several species, including *Campylobacter jejuni*, *C. coli* and *C. fetus*.

Epidemiology

• Zoonotic infection; reservoirs include cattle, poultry, wild birds and pets.
• Transmission via faecal–oral route often by consumption of contaminated undercooked or raw food (e.g. chicken, milk).
• Common cause of bacterial diarrhoea.

Morphology and identification

• Campylobacters have a characteristic morphology.
• They grow on selective media in a micro-aerophilic atmosphere plus 10% CO_2 at 42°C. Several selective media are available for culture of the organism, which contains charcoal and antibiotics to inhibit other bacteria. Colonies are colourless or grey with a metallic sheen.
• Campylobacters are oxidase and catalase posi-

tive. Further biochemical tests are used for speciation.

Pathogenicity

Campylobacter cell wall contains endotoxin. Cytopathic extracellular toxins and enterotoxins have also been demonstrated, but their exact role is unclear.

Associated infections

Acute colitis, rarely complicated by septicaemia, cholecystitis or reactive arthritis.

Laboratory diagnosis

• Culture of stool sample on selective media at 42°C for 42–72 h.
• Serology for serum antibodies to campylobacters is rarely used.

Treatment

Erythromycin or ciprofloxacin indicated only for severe diarrhoea.

Helicobacter

H. pylori

Definition

Curved Gram-negative bacilli; motile; non-sporing; non-capsulated.

Classification

H. pylori was originally classified as *Campylobacter pyloridis*.

Epidemiology

• *H. pylori* can survive attached to human gastric epithelial cells below the protective mucous layer.
• The organism is found most frequently in older patients. In the developed countries, it is present infrequently in children aged < 10 years (< 1%), but is found in about 50% of the population aged 50 years.
• Transmission is by the faecal–oral or oral–oral routes and is associated with close contact (intrafamilial) and poor sanitation. Humans are the main reservoir, and also cats.

Morphology and identification

• *H. pylori* grow at 37°C on enriched media (e.g. chocolate agar plates) in a microaerophilic atmosphere; colonies appear after 3–4 days.
• They possess four to six unipolar sheathed flagella (Plate 24); campylobacters have only a single polar unsheathed flagella; the test for urease activity is particularly important in laboratory identification of *H. pylori*.

Pathogenicity and associated infections

The most common cause of peptic ulceration. Various pathogenic mechanisms have been proposed for *H. pylori*: damages the mucous layer; cause inflammation with ulcer formation and may eventually result in cancer. Many with it are asymptomatic.

Laboratory diagnosis

• Culture of gastric biopsy specimens and serological tests for specific serum IgG antibodies.
• Gastric biopsy specimens may be examined directly for the presence of urease produced by *H. pylori* (growth of the organism is not required). Any urease present hydrolyses urea in a broth to produce ammonia; the rise in pH is detected by an indicator. This test is specific because *H. pylori* is unique in producing a urease with a high affinity and rate of activity. The test is simple, rapid and relatively inexpensive; however, some positives are missed.

• A breath test is also available. ^{13}C- or ^{14}C-labelled urea is given to the patient; if *H. pylori* is present, the urease splits the urea, producing labelled CO_2, which is detected in expired air by various methods.

Treatment

Clearance of *H. pylori* with two antibiotics (e.g. amoxicillin and clarithromycin) and an H_2 proton pump inhibitor (e.g. omeprazole) is effective in over 80% of cases. Combinations of other antimicrobial agents have also been used, including amoxicillin and metronidazole.

Vibrios

Vibrios are comma-shaped bacteria; the main species of medical importance are *Vibrio cholerae*, *V. parahaemolyticus*, *V. vulnificus* and *V. alginolyticus*.

V. cholerae

Definition

Gram negative; comma-shaped bacteria; aerobic and facultatively anaerobic; grow in alkaline conditions; motile; ferment carbohydrates; oxidase positive.

Classification

• *V. cholerae* strains are subdivided according to O-antigens. *V. cholerae* O1 is the cause of cholera. Other *V. cholerae* (non-O1) strains may occasionally cause diarrhoea.
• *V. cholerae* O1 has a number of biotypes including the 'classic' strain, which was the principal cause of cholera until the mid-1960s. The 'El Tor' biotype has been responsible for most cases of cholera over the last two decades. In the 1990s, O139 caused epidemics in south-east Asia.

Epidemiology

• Found in water contaminated with human faeces; no animal reservoirs.
• Important cause of severe infection in developing countries where potable water and sewage systems are poor.

• Transmitted via consumption of contaminated food or water.

Morphology and identification
• *V. cholerae* is characterized by an ability to grow in alkaline conditions (pH > 8.0). Alkaline broth is used to grow the organism selectively from faecal samples. A special selective medium (thiosulphate–citrate–bile salt–sucrose agar) is also used. *V. cholerae* forms characteristic yellow colonies on this medium.
• Specific slide agglutination reactions for 'O' antigens distinguish *V. cholerae* from non-cholera vibrios.
• Biochemical tests are used for confirmation.

Pathogenicity
V. cholerae produces an enterotoxin that acts on intestinal epithelial cells, stimulating adenylyl cyclase activity. This results in water and sodium ions passing into the gut lumen to produce profuse, watery diarrhoea (rice-water appearance).

Associated infections
Acute enteritis.

Laboratory diagnosis
Isolation of organism from faeces on selective media.

Prevention
Avoidance of contaminated food; oral vaccines suitable for travellers.

Treatment
Rehydration; ciprofloxacin shortens duration of illness.

Non-cholera vibrios

V. cholerae that do not agglutinate with O1 and O139 antisera are known as non-cholera vibrios. These organisms may also cause diarrhoea, which is normally much less severe than that from classic cholera.

V. parahaemolyticus

V. parahaemolyticus is a cause of food poisoning 12–18h after the ingestion of contaminated seafood, particularly oysters. The illness is characterized by vomiting and diarrhoea.

Treatment
By rehydration and a quinolone in severe cases.

Chapter 14

Treponema, Borrelia and *Leptospira*

Spirochaetes

The bacteria in the order Spirochaetales, commonly referred to as spirochaetes, are grouped together because of common morphological properties. They are thin (0.1–0.5 × 5.0–20.0 μm), helical, Gram-negative bacteria. Only a few spirochaetes can be cultured in vitro and their identification and diagnosis are based primarily on serological tests. The order Spirochaetales is divided into five genera, of which three, *Treponema*, *Borrelia* and *Leptospira*, cause human disease (Table 14.1).

Table 14.1 Spirochaetes and associated infections.

SPIROCHAETES	
Spirochaete	**Associated infections**
Treponema	
T. pallidum	Syphilis
T. pertenue	Yaws
T. carateum	Pinta
Borrelia	
B. recurrentis	Relapsing fever
B. burgdorferi	Lyme disease
Leptospira	Flu-like illness
	Meningitis
	Weil's disease

Treponema

Definition
Gram-negative spiral bacteria; motile; non-capsulated; non-sporing; non-culturable in vitro.

Classification
Treponema species that cause human disease are *Treponema pallidum* and *T. carateum*. *T. pallidum* causes syphilis and a subspecies, referred to as *T. pertenue*, causes yaws. *T. carateum* causes pinta.

T. pallidum

Epidemiology
T. pallidum is a strict human pathogen causing syphilis. Spread is by sexual contact; congenital infection can occur.

Morphology and identification
The organism is a thin, coiled spirochaete, 0.5 × 10.0 μm, which cannot be grown in vitro and grows only slowly in vivo. In clinical specimens, motile forms of the organism can be visualized by dark field microscopy or after specialized staining.

Associated infections
• Syphilis: this occurs worldwide and is a common sexually transmitted infection (STI) in the developed world; increasing in UK. The organism enters

the body via skin or mucous membrane abrasions. Localized multiplication occurs, resulting in inflammatory cell infiltration, followed by endarteritis. The infection is divided into three phases: primary, secondary and tertiary, and virtually any organ in the body can be invaded:

- Primary: involves development of the primary lesion, the chancre, at site of inoculation; usually heals spontaneously in about 6 weeks.
- Secondary (6–8 weeks from infection): dissemination stage, via bloodstream to various tissues; when this subsides the infection enters a latent phase.
- Tertiary (develops > 3 years later): occurs in about one-third of untreated patients; involves aorta or arteries of central nervous system (CNS).

- Congenital syphilis: transmission to the fetus may result in intrauterine death or congenital abnormalities.

Laboratory diagnosis

As *T. pallidum* cannot be grown in vitro on standard laboratory media, diagnosis is based on microscopy and serology:

Microscopy

Primary, secondary and congenital syphilis can be diagnosed by dark field examination of fresh material from skin lesions.

Serology

Diagnosis of syphilis in most patients is based on serological tests. Two types of test are employed: non-specific and specific treponemal tests. The interpretation of serological tests for syphilis in relation to disease activity is discussed later.

1 Non-specific antibody tests: these rapid tests determine the presence of IgM and IgG antibodies, which develop against lipids released from damaged *T. pallidum* during the early stages of the disease. The antigen used in these assays, 'cardiolipin', is extracted from beef heart; cardiolipin binds antibodies to *T. pallidum* lipids. The assays are non-specific but are useful in the diagnosis of active syphilis. The tests commonly used are the Venereal Disease Research Laboratory (VDRL) test

and the rapid plasma reagent (RPR) test. Both assays are based on the agglutination of cardiolipin antigen by the patient's serum. As the assays are non-specific, false-positive results are common and occur in other conditions (e.g. leprosy, tuberculosis, viral infections, malaria, rheumatoid arthritis); results need confirmation by specific serological tests.

2 Specific antibody tests are based on *T. pallidum* antigens and are used to confirm non-specific screening tests. Commonly used tests include the fluorescent treponemal antibody (FTA) test and the particle agglutination test e.g. *T. pallidum* haemagglutination assay (TPHA), in which erythrocytes coated with *T. pallidum* antigen are agglutinated by serum from patients with antibodies to *T. pallidum*. Positive results may reflect old disease because tests remain positive after treatment. False-positive results may occur in other treponemal infections, e.g. yaws, pinta.

Specific tests to detect IgM antibodies are used to diagnose congenital infections.

Treatment

Penicillin.

T. carateum

Pinta is caused by *T. carateum* and is found primarily in Central and South America. It occurs in all age groups. After a 1- to 30-week incubation period, small papules develop on the skin. These can enlarge, last for several years and result in hypopigmented lesions. Spread is via direct contact or insect vectors. Diagnosis and treatment are as for syphilis.

T. pertenue

T. pertenue is the causative agent of yaws. This is primarily a skin disease; destructive lesions (granulomata) of the skin, lymph nodes and bone can occur. Yaws is found primarily in South America and Central Africa.

Spread is by direct contact with infected skin lesions. Diagnostic procedures and treatment are as for syphilis.

Borrelia

Borrelia species are associated with two important human infections: relapsing fever (primarily *B. recurrentis*) and Lyme disease (*B. burgdorferi*).

Definition

Gram-negative, spiral bacteria; motile; non-capsulated; non-sporing; require specialized media for culture; anaerobic or microaerophilic.

B. recurrentis, B. duttoni and other *Borrelia* species

Relapsing fever

Occurs worldwide; characterized by episodes of fever and septicaemia, separated by periods when the patient is apyrexial:
• Epidemic relapsing fever is caused by *B. recurrentis* and is spread via the human body louse.
• Endemic relapsing fever is caused by many species of *Borrelia* and is spread from rodents via infected ticks. The insect can infect their young transovarially (i.e. via the egg in the ovary), maintaining an endemic reservoir.
The characteristic relapsing picture is related to the ability of *Borrelia* species to vary their antigenic structure and avoid specific host antibodies.

B. recurrentis can be identified in stained blood smears, taken when a patient is pyrexial. Isolation may be successful with enriched media in anaerobic conditions. Serological tests are not often helpful as a result of antigen variation. Tetracycline is the treatment of choice.

B. burgdorferi

B. burgdorferi is the cause of Lyme disease. The organism is transmitted to humans by the ixodes tick from animal hosts (deer and rodents). Infection spreads mainly in forested areas and is characterized initially by a skin lesion at the site of the tick bite, but later involves joints, the heart and the CNS.

Diagnosis is by serological tests for IgM and IgG. Cephalosporin (cefotaxime) or tetracycline is an effective antibiotic. Prevention is by avoidance of tick bites.

Leptospira

Definition

Gram-negative, coiled bacteria; non-capsulated; non-sporing; motile; aerobes that grow slowly in enriched media.

Classification

The genus *Leptospira* has two species: *L. interrogans* and *L. biflexa*. *L. interrogans* causes disease in humans, whereas *L. biflexa* is a free-living saprophyte. There are over 100 serotypes of *L. interrogans* but only a few are associated with human disease, namely *icterohaemorrhagiae, canicola, hardjo* and *pomona*.

Epidemiology

• Leptospirosis occurs worldwide. Animals act as a reservoir of human infection, e.g. rats (*icterohaemorrhagiae*), dogs (*canicola*), cows (*hardjo*) and pigs (*pomona*). About 50 cases/year in the UK.
• Human infections occur mainly after exposure to water contaminated with infected urine. At-risk groups include sewer workers, farmers and those involved in watersports.

Morphology and identification

Leptospires are motile organisms with two periplasmic flagella, each anchored at opposite ends of the organism. They can be grown on specially defined media enriched with serum, although incubation for up to 2 weeks may be required.

Associated infections

Infections range from a mild flu-like illness, uveitis and meningitis to Weil's disease (with liver and renal damage).

Laboratory diagnosis

Blood or urine may be cultured in specialized media but takes weeks. Diagnosis is often made serologically by detecting the presence of leptospiral antibodies.

Treatment and prevention

• Treatment is with penicillin or tetracycline.
• Rodent control measures help to prevent infection. Groups at risk should cover cuts and abrasions.

Chapter 15

Gram-negative anaerobic bacteria

The Gram-negative anaerobes include the genera *Bacteroides*, *Fusobacterium* and *Leptotricia*. Unlike the Gram-positive anaerobic clostridia, they do not form spores. Gram-negative anaerobes form a major component of the normal flora of humans.

Definition
Gram-negative bacilli; non-motile; non-sporing; strict anaerobes; grow on complex media; some species capsulate.

Bacteroides

The genus *Bacteroides* contains a large number of species, the most important being *B. fragilis*.

Epidemiology
Bacteroides species make up the largest component of the bacterial flora of the human intestine, and are an important part of the normal flora of the mouth and vagina. They protect against colonization and infection by more pathogenic bacteria; they can also cause infections.

Morphology and identification
Gram-negative bacilli; occasionally form short filaments. Some species have a polysaccharide capsule. They are strictly anaerobic organisms that grow on complex media; visible colonies appear only after 48 h of incubation. Some species produce a pigment that results in black colonies. (*Note*: taxonomic changes have placed black-pigmented *Bacteroides* in a separate genus, *Prevotella*; see Plate 25.)

Associated infections
These include: intra-abdominal sepsis (wound infections, abscesses, peritonitis) often in mixed infections with coliforms; abscesses in other organs, particularly brain (again, often mixed infections); oral infections, e.g. acute ulcerative gingivitis; and aspiration pneumonia and lung abscesses.

Prevention
• Postoperative, intra-abdominal infections can be prevented by good surgical technique and prophylactic antibiotics.

Treatment
• Metronidazole and clindamycin are effective antibiotics against *Bacteroides* species; most species are sensitive to penicillin but some, particularly *B. fragilis*, are resistant because of β-lactamase production.

Fusobacterium

Fusobacterium species are spindle-shaped, Gram-negative bacilli with pointed ends. They are strict anaerobes and are often nutritionally demanding. The genus is classified into various species, *F.*

necophorum and *F. nucleatum* being the most important in human infections.

Fusobacteria are commensals of the oral cavity. They are one of the causative organisms, along with other anaerobes, of acute ulcerative gingivitis (primarily *F. nucleatum*) and other oral infections. *F. necophorum* may occasionally cause septicaemia with metastatic abscesses.

Leptotricia

Leptotricia species are long, slightly curved rods with pointed ends. The genus consists of a single species, *L. buccalis*, which is part of the normal oral flora and may be associated with oral infections such as Vincent's angina.

Chapter 16

Mycobacteria

Definition
Acid-fast bacilli; obligate aerobes; non-capsulated; non-motile; grow slowly on specialized media; cell walls have large lipid content.

Classification
- The genus includes: *Mycobacterium tuberculosis* and *M. bovis* (tuberculosis); *M. leprae* (leprosy); also the atypical mycobacteria, including *M. avium-intracellulare*, *M. kansasii* and *M. marinum*.
- Mycobacteria stain poorly by Gram stain as a result of cell-wall mycolic acid. They are recognized by the Ziehl–Neelsen (ZN) stain (Plate 26).
- Mycobacteria grow slowly; visible colonies appear after 1–12 weeks depending on the species (Plate 27).
- Classification is based on culture characteristics, including nutritional requirements, rate of growth, pigmentation and biochemical properties.

M. tuberculosis and *M. bovis*

M. tuberculosis, commonly called the tubercle bacillus (primary host humans), and *M. bovis* (primary host cattle) are similar species of mycobacteria and cause similar infections.

Epidemiology
- Tuberculosis (TB) is the most common infectious disease worldwide.

- Increasing in incidence related to HIV, mostly in Asia and Africa.
- *M. tuberculosis* infections are spread usually by inhalation of droplets, and rarely by ingestion. Incubation period is 4–16 weeks. Human infection with *M. bovis* is acquired via contaminated milk. TB is highly infectious and outbreaks may occur.
- Mycobacteria are able to survive for long periods in the environment because they withstand drying.

Pathogenicity
- *M. tuberculosis*, in common with other mycobacteria, is an intracellular pathogen; its survival within macrophages is related to its ability to prevent phagosome–lysosome fusion.
- The host response to mycobacterial infection is via the cell-mediated immune system and results in the formation of granulomata. After inhalation, a pleural lesion (Ghon focus) develops; infection spreads to the hilum and lymph nodes (primary complex) with granuloma formation. In most cases the complex heals. In 10% of patients, infection spreads to other sites. Some sites become dormant and may reactivate years later.

Associated infections
- Pulmonary TB (75%)
- Extrapulmonary TB, which includes: meningitis; osteomyelitis; miliary TB (multisystem infection—associated with night sweats, fever and weight

loss); cervical and mesenteric lymphadenopathy; abdominal and renal infection.

- *M. bovis* infection is typically localized to bone marrow and cervical or mesenteric lymph nodes.

Laboratory diagnosis
- Microscopy of smears of sputum or urine or tissue by ZN stain and appearance as thin bacilli with beads (positive in < 60% pulmonary; < 25% extrapulmonary). A fluorescent rhodamine–auramine dye can also be used.
- Culture on special media for up to 12 weeks, e.g. Lowenstein–Jensen medium, which contains egg yolk, glycerol and mineral acids plus inhibitors, such as malachite green, to reduce growth of other bacteria. Specimens, e.g. sputum, contaminated with normal flora, are pre-treated with acid to reduce microbial contamination. Biochemical tests, pigment production or DNA probes are used to confirm identification. Antibiotic sensitivity can be obtained in about 2 weeks using radiometric and non-radiometric automated systems and DNA probes.
- Immunological tests to detect *M. tuberculosis* (e.g. Mantoux test): these are based on the inoculation of purified protein derivative (PPD), derived from tubercle culture filtrate, into the patient's skin to demonstrate cell-mediated immunity to *M. tuberculosis*. In some countries, the widespread use of vaccination, or the frequency of previous exposure to tuberculosis, decreases the usefulness of skin testing.

Treatment and prevention
- Treatment is with combinations of antimycobacterial drugs (e.g. rifampicin, isoniazid, pyrazinamide and ethambutol) for 2 months (initial phase), followed by 4 months (continuous phase) of rifampicin and isoniazid (longer course of treatment required for meningitis). Multiply drug resistant strains of *M. tuberculosis* (MDRTB) are now emerging (< 5% in the UK).
- For prevention of *M. tuberculosis* infection different strategies have been tried, including widespread immunization with a live attenuated vaccine (bacille Calmette–Guérin [BCG]), and isolation and prompt treatment of cases combined with BCG as appropriate, and chemoprophylaxis of close contacts.

- *M. bovis* infection is prevented by destruction of tuberculin-positive cattle and milk pasteurization.

M. leprae

M. leprae (Hansen's bacillus) causes leprosy.

Epidemiology
- There are over 10 million cases of leprosy worldwide, mainly in Asia, Africa and South America. There are no animal reservoirs.
- Transmission follows prolonged exposure to shedders of the bacilli via respiratory secretions or ulcer discharges. The incubation period is 2–10 years; without prophylaxis up to 10% develop the disease.

Morphology and identification
M. leprae is an acid-fast bacillus. It can be grown in the footpads of mice or armadillos, from tissue culture and on artificial media.

Associated infections
There are two major types of leprosy: lepromatous and tuberculoid, with various intermediate stages:
- Lepromatous: progressive infection, resulting in nodular skin lesions (granulomata) and nerve involvement; associated with a poor prognosis.
- Tuberculoid: a more benign, non-progressive form that involves macular skin lesions and severe asymptomatic nerve involvement. Spontaneous healing usually results after tissue and nerve destruction.

Laboratory diagnosis
Microscopy of scrapings from skin or nasal mucosa, or skin biopsies, examined by ZN staining.

Treatment and control
- Treatment: combination of dapsone plus rifampicin or clofazimine.
- Chemoprophylaxis is needed for close contacts of infected individuals, and is particularly important for young children in contact with adults with leprosy.

Table 16.1 Infections associated with atypical mycobacteria.

ATYPICAL MYCOBACTERIA	
Mycobacterium species	Associated infections
M. ulcerans	Skin ulcers (tropics)
M. kansasii	Chest disease
M. marinum	Granulomatous skin lesions associated with cleaning fish tanks and swimming pools
M. chelonei	Cutaneous abscesses; occasionally disseminated infection in immunocompromised individuals
M. avium-intracellulare	Cervical lymphadenopathy in children; chest infection and invasive infections in AIDS patients

Atypical mycobacteria

These grow at a variety of temperatures; some are rapid growing (3–4 days), whereas others are slower (> 8 weeks). Some produce pigmented colonies in light (photochromogens) or in light and dark (scotochromogens).

They are transmitted to humans primarily from environmental or animal sources. Associated infections are listed in Table 16.1. Some species, e.g. *M. avium-intracellulare*, are resistant to first-line anti-mycobacterial agents.

Chapter 17

Chlamydiae, Rickettsiaceae, mycoplasmas and L-forms

Chlamydiae

Definition
Obligate intracellular bacteria; normally grown in tissue cultures.

Classification
The genus *Chlamydia* is divided into three species: *Chlamydia psittaci*; *C. trachomatis* and *C. pneumoniae*. Classification is based on antigens, morphology of intracellular inclusions and disease patterns (Table 17.1).

Morphology and identification
All chlamydiae share a common group antigen but may be distinguished by species-specific antigens. They multiply in the cytoplasm of host cells after a well-defined developmental cycle. Chlamydiae have two distinct morphological forms:

1 The elementary body (EB): an infectious extracellular particle (300–400 nm).
2 The reticulate body (RB): an intracellular non infectious particle (800–1000 nm).

Life cycle
• EBs bind to specific host cell receptors and enter by endocytosis. Target cells include conjunctival, urethral, rectal and endocervical epithelial cells.
• Intracellular EBs remain within phagosomes and replicate. Lysosomal fusion with the phagosome-

Table 17.1 Classification and common infections of *Chlamydia*.

CHLAMYDIA				
Species	**Serotypes**	**Natural host**	**Common infections**	**Transmission**
C. trachomatis	A–C	Humans	Trachoma/conjunctivitis	Hands/eyes/flies
C. trachomatis	D–K	Humans	Cervicitis	
			Conjunctivitis	
			Urethritis	Sexual/perinatal
			Proctitis	
			Pneumonia	
C. trachomatis	L1, L2, L3	Humans	Lymphogranuloma venereum; genital ulcers	Sexual
C. psittaci	Many	Birds and some mammals	Pneumonia	Aerosol
C. pneumoniae	1	Humans	Acute respiratory infections	Aerosol

containing EBs is inhibited, probably by chlamydial cell-wall components.

- After about 8 h, EBs become metabolically active and form RBs that synthesize DNA, RNA and protein, utilizing host cell adenosine triphosphate (ATP) as an energy source.
- RBs undergo multiple division and the phagosome becomes an 'inclusion body' (visible by light microscopy).
- After about 24 h, the RBs reorganize into smaller EBs and, after a further 24–48 h, the host cell lyses and infective EBs are released.

Pathogenicity
Chlamydiae are intracellular parasites avoiding cellular defences by inhibiting phagosome–lysosome fusion.

C. trachomatis

Classification
C. trachomatis can be divided into different serotypes ('serovars'): serotypes A–C, D–K, and L1, L2 and L3.

Epidemiology
C. trachomatis serotypes D–K are found worldwide, whereas the lymphogranuloma venereum (LGV) serotypes occur primarily in Africa, Asia and South America. The majority of C. trachomatis infections are genital and sexually acquired. Asymptomatic infections can occur in women. Eye infections in adults probably result from oculogenital contact. Similarly, neonates can be infected during birth from an infected mother.

Pathogenesis
C. trachomatis gains access via disrupted mucous membranes and can infect mucosal columnar or transitional epithelial cells, resulting in severe inflammation. In LGV, the lymph nodes, which drain the site of the primary infection, are involved.

Associated infections
- Trachoma: keratoconjunctivitis caused by serotypes A–C; the leading cause of preventable blindness in the world.

- Adult inclusion conjunctivitis: a milder disease than trachoma; caused by serotypes D–K; associated with genital infections.
- Neonatal conjunctivitis: C. trachomatis can be acquired at birth from an infected mother; associated with serotypes D–K.
- Neonatal pneumonitis: presents 3–12 weeks after birth; often associated with neonatal conjunctivitis; acquired through maternal transmission during birth.
- LGV: a sexually transmitted infection (STI), with genital ulcers and suppurative inguinal adenitis. It is common in tropical climates and associated with serotypes L1–L3.
- Urogenital infections (urethritis and cervicitis): associated with serotypes D–K; the most common cause of STIs in the developed world. Complications include pelvic inflammatory disease and perihepatitis in women, and epididymitis and Reiter's syndrome in men.

Laboratory diagnosis
- Culture: swabs from the affected site, collected in a special transport medium, are inoculated into tissue culture. Growth of C. trachomatis is recognized by Giemsa staining, showing intracellular bodies, or by immunofluorescence staining with specific antisera.
- Direct antigen detection can be made by immunofluorescence (Plate 28) or enzyme-linked immunosorbent assay (ELISA) methods. Both techniques utilize labelled specific antibodies to C. trachomatis.
- Serology: antibodies to C. trachomatis may be detected by complement fixation or micro-immunofluorescence tests. Serological testing has limited value because it is difficult to distinguish between past and current infections.
- Nucleic acid amplification tests to detect either chlamydial DNA or RNA, using polymerase chain reaction (PCR) or ligase chain reaction, or chlamydial ribosomal RNA, using transcription-mediated amplification, are now available. These are performed on urine and vaginal swabs. They are more sensitive than culture, non-invasive and replacing the above tests.

Treatment
• Urogenital infection: tetracycline or azithromycin.
• Trachoma: for individual or sporadic cases, tetracycline (erythromycin for children, breast-feeding or pregnant women). Eyelid deformities may need surgery.
• Neonatal conjunctivitis: erythromycin.

C. psittaci

C. psittaci causes infection in psittacine birds (e.g. parrots and budgerigars) and may be transmitted to humans.

Epidemiology
Transmission to humans is via inhalation, usually of dried bird guano. Pet shop workers, bird fanciers and poultry farm workers are at particular risk of developing psittacosis.

Associated infections
Flu-like illness and pneumonia.

Laboratory diagnosis
By serology, including immunofluorescence or complement fixation test. Culture of *C. psittaci* is also possible but is performed only in laboratories with high-level containment facilities.

Treatment
Tetracycline is the drug of choice; erythromycin is an alternative.

Prevention
Treat infected birds; quarantine.

C. pneumoniae

This was originally designated TWAR (Taiwan-associated respiratory disease) agent, but is now called *C. pneumoniae* and is a significant cause of atypical pneumonia.

Infection is spread via direct human contact, with no apparent animal reservoir. It is associated with respiratory infections, including pharyngitis, sinusitis and pneumonia (which can be severe, particularly in elderly people). Diagnosis is by serology for antibodies against *C. pneumoniae* (detection of *C. pneumoniae* DNA by PCR is being developed); treatment includes either tetracycline or erythromycin.

Rickettsiaceae

Rickettsiaceae cause a number of important human infections including typhus, the related spotted fevers and Q-fever.

Definition
Small, Gram-negative bacilli (0.2 × 0.7 μm diameter); obligate intracellular pathogens; utilize ATP from host cell; grow only in tissue culture.

Classification
Rickettsiaceae contain two important genera: *Rickettsia* and *Coxiella*.

Rickettsia

The genus contains a number of species that cause human infection (Table 17.2). (Similar related organisms in genus *Orientia*.)

Epidemiology
Rickettsial infections are zoonoses with a variety of animal reservoirs and insect vectors.

Associated infections
Typhus and spotted fevers.

Laboratory diagnosis
This is by serology. The Weil–Felix test detects cross-reacting antibodies to rickettsiae which agglutinate certain strains of *Proteus* (OX-19 and OX-2). The test is non-specific and has now been superseded by more specific serological tests based on purified rickettsial antigens (immunofluorescence assays and ELISA).

Treatment and prevention
Treatment is with tetracycline or chloramphenicol. Infection can be prevented by avoidance of the

Table 17.2 Vectors and infections associated with *Rickettsia*.

RICKETTSIA			
Species	Principal host/reservoir	Vector	Disease
R. typhi	Rats	Fleas	Murine typhus
R. prowazeki	Humans/squirrels	Lice	Louse-bound typhus
R. tsutsugamushi	Rats	Mites	Scrub typhus
R. akari	Mice	Mites	Rickettsial pox
R. rickettsii[a]	Dogs	Ticks	Rocky mountain spotted fever

[a] Tick-borne spotted fevers in areas of the world outside the USA carry a variety of names and are caused by rickettsiae very similar to *R. rickettsii* (e.g. Boutonneuse fever caused by *R. conorii* occurs in the Mediterranean region).

various vectors. A vaccine to *R. prowakzekii* is available.

Coxiella

A pleomorphic coccobacillus with a Gram-negative cell wall (0.4–0.7 µm). Spores are produced prolonging survival. This is an animal pathogen (commonly cattle, sheep; goats) that occasionally infects humans via tick bites or inhalation of dried, contaminated matter. At-risk groups include farmers, vets and abattoir workers. Urine, faeces, milk and birth products may be a source. Outbreaks can occur.

Coxiella burnetii causes Q-fever (non-specific illness of fever, severe headache, chills, severe malaise, myalgia and later pneumonitis), endocarditis, hepatitis and neurological infections. Contact with animals gives a clue to the diagnosis. Laboratory diagnosis is by serology (complement fixation, ELISA and immunofluorescence tests). Treat with tetracycline. For endocarditis, treatment is prolonged and valve replacement may be needed. Vaccine is being developed for prevention.

Mycoplasma and Ureaplasma

Definition

Small (< 1.0 µm), Gram-negative organisms; pleomorphic with no cell wall; grow slowly on enriched media; aerobic or facultatively anaerobic.

Classification

The family Mycoplasmataceae consists of two genera: *Mycoplasma* (69 species) and *Ureaplasma* (two species). Only a few species have been identified as human pathogens, including: *Mycoplasma pneumoniae*, *M. hominis*, *M. genitalium*, *M. fermentans* and *Ureaplasma urealyticum*.

Epidemiology

M. pneumoniae infection is common worldwide and may cause up to 30% of acute lower respiratory tract infections; epidemics can occur. Immunity to *M. pneumoniae* is short-lived. Infection is spread by nasal secretions. *M. hominis* and *U. urealyticum* colonize the genitourinary tracts.

Associated infections
• *M. pneumoniae*: pharyngitis; community-acquired pneumonia, extrapulmonary involvement including skin, heart and central nervous system (CNS).
• *U. urealyticum/M. hominis/M. genitalium*: urethritis; possible cause of pelvic inflammatory disease and opportunist infection in immunocompromised individuals (arthritis, meningitis).

Laboratory diagnosis
• Culture: mycoplasmas and ureaplasmas can be grown on enriched media; penicillin is often added to inhibit other organisms. Although *M. pneumoniae* can be isolated from sputum after incubation for up to 3 weeks, diagnosis is normally by serology. *M. hominis* grows after about 4 days, producing colonies with a 'fried egg' appearance. *U. urealyticum* requires urea for growth and forms small colonies.

• Serology: *M. pneumoniae* infections can be diagnosed by serological tests (complement fixation test or enzyme immunoassays) for IgG (fourfold rise in titres is indicative of current infection) or IgM.

Treatment
Macrolides (e.g. erythromycin) or tetracycline.

L-forms

L-forms, named after the Lister Institute where they were first reported, are cell-wall-deficient forms of bacteria. L-forms need to be maintained on osmotically stabilized media to prevent lysis. They are able to multiply and their colonial morphology is similar to the 'fried egg' appearance of the mycoplasmas. A few L-forms are also able to re-form their cell walls and revert back to parent vegetative state with a complete cell wall. These are referred to as unstable L-forms. Other L-forms cannot produce new cell walls and are referred to as stable L-forms. It has been suggested that mycoplasmas are stable L-forms.

L-forms have been associated with renal infections but their clinical significance is unclear.

Chapter 18

Basic virology

Virus structure (Figure 18.1)

Size and shape

Viruses range in size from 20 to 300 nm in diameter and many have unique shapes that aid identification. All viruses have one of two shapes (symmetry): icosahedral with two-, three- or fivefold axis of symmetry, or helical symmetry.

Genome

Viruses contain either DNA or RNA. Nucleic acids represent the main component of the virus core and are associated with core proteins. Viral nucleic acids range from 1.5×10^6 Da (parvoviruses) to 200×10^6 Da (poxviruses) and may be single stranded or double stranded, circular or linear, a single molecule or in segments.

Capsid

A protein coat encloses the genome and core proteins, consisting of capsomeres (capsid subunits).

Envelope

Lipid bilayer membrane surrounding the nucleocapsid of some viruses (enveloped viruses); the envelope carries glycoproteins, which form projections or spikes.

Virion

The complete infectious virus particle; may lack nucleic acids (empty particles) or carry defective genomes (defective particles), which can interfere with normal replication (defective interfering particles).

Virus cultivation

Viruses are obligate, intracellular parasites and thus can replicate only in living cells. Viruses utilize the host cell metabolism to assist in the synthesis of viral proteins and progeny virions; the host cell range of viruses may be narrow or wide.

For diagnostic purposes, most viruses are grown in cell cultures, either secondary or continuous cell lines; the use of embryonic eggs and laboratory animals for virus culture is reserved for specialized investigations. Cell culture lines for virus cultivation are often derived from tumour tissues, which can be used indefinitely.

Virus replication in cell cultures may be detected by:
- *Cytopathic effect (CPE)*: some viruses can be recognized by their effect on cell architecture, e.g. necrosis, lysis, the presence of inclusion bodies or the formation of multinucleated cells.
- Haemadsorption: viruses expressing haemagglutinins on the cell surface may be recognized by absorption of red blood cells to infected cells.
- Immunofluorescence: the appearance of virus-coded proteins on the surface, or in the nucleus or cytoplasm of infected cells may be detected by im-

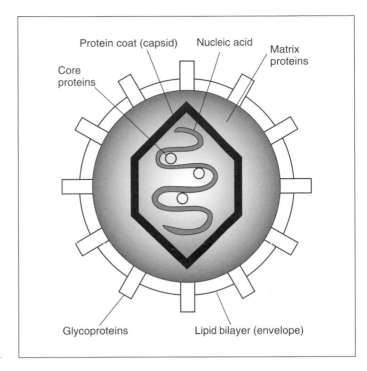

Figure 18.1 An enveloped virus.

munofluorescence techniques using virus protein-specific antibody.

Viral infection of host cells

Virus replication in host cells involves the following steps:

1 Attachment (adsorption) of the viral nucleocapsid (naked viruses) or of virus envelope components (enveloped viruses) to cell surface molecules (receptors); this involves specific interaction between viral glycoproteins (e.g. the haemagglutinin of influenza virus) and host cell surface components (e.g. *N*-acetylneuraminic acid for influenza virus). Many viruses have highly specific receptors.

2 Penetration of the virus into the host cell takes place (often by receptor-mediated endocytosis).

3 Uncoating follows, which involves the proteolytic removal of viral protein coat and liberation of nucleic acid and attached core proteins.

4 In order to direct the host cell ribosome to produce viral proteins (core, capsid), virus-specific mRNA must be produced. The mechanisms for virus-specific mRNA production depend on the viral genome type:

(a) RNA or DNA

(b) single or double stranded

(c) positive sense (base sequence configured as required for translation = mRNA sense) or negative sense (base sequence requires transcription). Examples are shown in Figure 18.2. The mechanisms for replication of viral nucleic acid also depend on viral genome type; examples are given in Figure 18.3.

5 Morphogenesis and maturation occur with assembly of components (nucleic acid, proteins) to form subviral particles (pre-core, core particles) and viral particles (virions, empty particles, DI particles).

6 Release of the virus is by bursting of infected cells (lysis) or by budding through the plasma membrane (host cell does not necessarily lose viability, so it may shed viral particles for extended periods).

With some viruses, e.g. hepatitis B, the host cell remains viable and continues to release virus

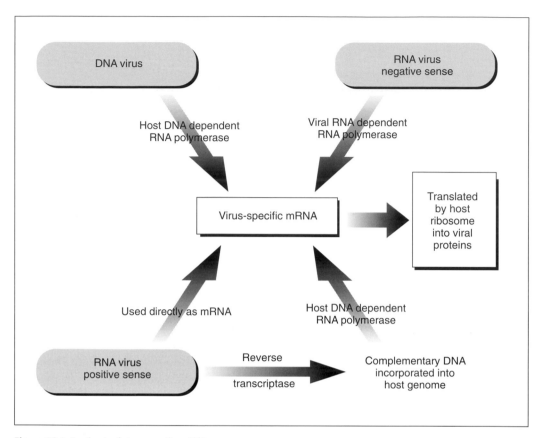

Figure 18.2 Synthesis of virus-specific mRNA.

Table 18.1 Viral classification according to disease or organ system involved.

VIRAL CLASSIFICATION	
Organ system involved (disease)	**Examples of clinically important viruses**
Systemic infections	Measles virus, rubella virus, chickenpox virus, enteroviruses, retroviruses
Central nervous system	Polio- and other enteroviruses; rabies virus, arthropod-borne viruses, herpes simplex virus, measles virus, mumps virus, retroviruses, cytomegalovirus (CMV)
Respiratory tract (common cold, tracheitis, bronchitis, bronchiolitis, pneumonia)	Influenzavirus, parainfluenzavirus, respiratory syncytial virus, adenovirus (enteroviruses), CMV
Eye (conjunctivitis, retinitis)	Herpes simplex virus, adenovirus, CMV
Skin and mucous membranes (rash, warts)	Herpes simplex virus, papillomavirus, chickenpox virus, measles virus, rubella virus, parvovirus
Liver	Hepatitis viruses A–G, yellow fever virus, herpesviruses, rubella virus
Salivary glands	Mumps virus, CMV
Gastrointestinal tract	Rotaviruses, adenoviruses, astroviruses, caliciviruses

Note: this classification is clinically oriented and not virus systematic. The same clinical symptoms can be caused by numerous viruses, and one virus can cause different clinical syndromes.

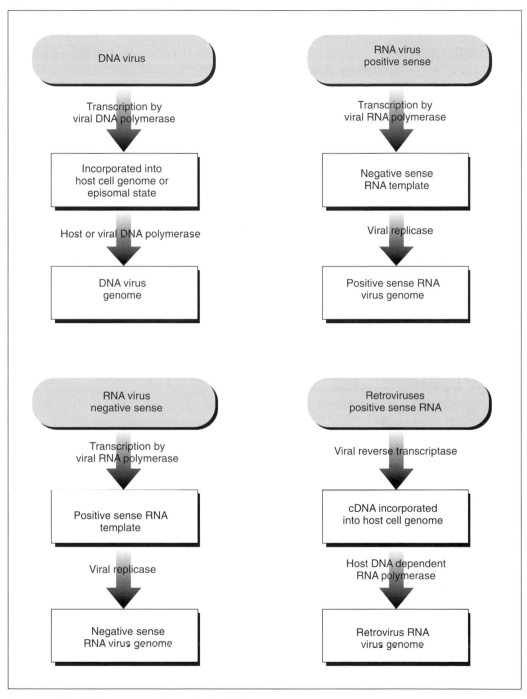

Figure 18.3 Replication strategies of some DNA and RNA viruses.

Table 18.2 Classification of virus families according to nucleic acid and virion structure.

CLASSIFICATION OF VIRUS FAMILIES			
Nucleic acid type	**Symmetry of nucleocapsid**	**Envelope**	**Virus family**
DNA SS linear	Icosahedral	−	*Parvoviridae*
DNA DS circular	Icosahedral	−	*Papovaviridae*
DNA DS linear	Icosahedral	−	*Adenoviridae*
DNA DS linear	Icosahedral	+	*Herpesviridae*
DNA DS linear	Complex	+	*Poxviridae*
DNA DS circular	Icosahedral	+	*Hepadnaviridae*
RNA SS linear	Icosahedral	−	*Picornaviridae*
RNA SS linear	Icosahedral	−	*Caliciviridae*
RNA SS linear	Icosahedral	+	*Togaviridae*
RNA DS	Icosahedral	−	*Reoviridae*
RNA SS	Complex	+	*Flaviviridae*
RNA SS	Complex	+	*Arenaviridae*
RNA SS	Icosahedral	+	*Coronaviridae*
RNA SS	Icosahedral	+	*Retroviridae*
RNA SS	Helical	+	*Orthomyxoviridae*
RNA SS	Helical	+	*Bunyaviridae*
RNA SS	Helical	+	*Paramyxoviridae*
RNA SS	Helical	+	*Rhabdoviridiae*

DS, double stranded; SS, single stranded; −, absent; +, present.

particles or subviral antigens at a slow rate. These persistent infections act as a continuing source of new infectious viruses.

During latent infections the virus does not undergo replication; the viral nucleic acid may remain in the host-cell cytoplasm (e.g. herpesvirus) or become incorporated into the host genome (e.g. human immunodeficiency virus [HIV]). A trigger is required to recommence viral nucleic acid replication, transcription and translation.

Virus classification

Viruses can be classified according to the following:
1 Disease or organ system involved (Table 18.1)

2 Nucleic acid type/virion structure (virus families; Table 18.2)
3 Replication strategy (see Figure 18.3).

Detection of virus infections

1 Direct (detection of virus particles, viral antigen, viral lesions or nucleic acid) by: microscopy; electron microscopy; particle agglutination; immunofluorescence; serology; gene probes including; polymerase chain reaction (PCR) and reversed transcription PCR (RT-PCR).
2 Indirect (detection of virus-specific host response) by serology (including complement fixation test, haemagglutination inhibition test, or enzyme-linked immunosorbent assay [ELISA]).

Chapter 19

Major virus groups

Herpesviruses

The herpesviruses are a family of enveloped DNA viruses that characteristically cause latent infections. Following primary infection, viral DNA lies dormant in various tissues and may be reactivated. Herpesviruses include:

- herpes simplex virus (HSV, type 1 and type 2)
- varicella-zoster virus (VZV)
- cytomegalovirus (CMV)
- Epstein–Barr virus (EBV)
- human herpesviruses (HHV), types 6–8.

Herpes simplex virus (HSV)

Virus
Enveloped DNA virus, 100–300 nm in diameter. There are two virus types: HSV-1 and HSV-2. They share 50% sequence identity and can be distinguished serologically.

Epidemiology
HSV has worldwide distribution. Antibodies to HSV-1 are acquired early in childhood and 90% of adults show serological evidence of past infection. Antibodies to HSV-2 are acquired mainly during adulthood between 14 and 40 years of age.

Pathogenesis
Spread is by direct contact (sexual activity is the main route of HSV-2 transmission). The incubation period is variable (2–12 days). There are local IgA, IgM and IgG responses, but the virus avoids elimination and its DNA becomes dormant in sensory ganglia. Reactivation is associated with various stimuli (e.g. stress, immunocompromised status, bacterial infection). Infections are primary, recurrent (by the same type) or initial (patient already infected with one type and has first infection with opposite type).

Associated diseases
- Primary disease: oropharyngeal disease (HSV-1, less commonly HSV-2); herpes genitalis (HSV-2, less commonly HSV-1); keratoconjunctivitis, skin infections.
- Recurrent disease: herpes labialis (cold sores); recurrent herpes genitalis; keratoconjunctivitis.
- Invasive infections can lead to: pneumonitis, oesophagitis and encephalitis, particularly in neonates and immunocompromised patients.

Diagnosis
By electron microscopy of vesicle fluid; direct fluorescent antibody staining of vesicle material; isolation in tissue; serology (e.g. enzyme-linked immunosorbent assay [ELISA]); nucleic acid detection.

Treatment
Aciclovir, valaciclovir, famciclovir, foscarnet, cidofovir (Chapter 31).

Varicella-zoster virus (VZV)

Virus
Enveloped DNA virus, 150–200 nm in diameter.

Epidemiology
VZV has a worldwide distribution. Infections acquired early in childhood (90% of adults are seropositive).

Pathogenesis
Spread is by droplets or direct contact with vesicle fluid. The incubation period is 2 weeks with a range of 7–23 days. The virus enters via the respiratory tract followed by viraemia and a generalized rash (chickenpox). It may become latent in the sensory ganglia; a variety of stimuli (e.g. immunosuppression) results in reactivation with vesicles in various dermatomes (shingles).

Associated diseases
Primary infection is chickenpox, and recurrent infection shingles. Complications: disseminated VZV infection in immunocompromised individuals and neonates, encephalitis, congenital varicella syndrome, ophthalmic zoster.

Diagnosis
This is made by: electron microscopy of vesicle fluid; direct fluorescent antibody staining of vesicle material; isolation in tissue culture (fibroblasts); serology, nucleic acid detection.

Treatment
Aciclovir, valaciclovir, famciclovir, foscarnet.

Prevention
• Passive: zoster immunoglobulin (ZIG) for protection of seronegative contacts at risk of VZV infection, i.e. pregnant women and immunocompromised patients.
• Active: live attenuated vaccine.

Cytomegalovirus (CMV)

Virus
Enveloped DNA virus, 200 nm in diameter.

Epidemiology
CMV has a worldwide distribution. In the developing world > 90% of adults are seropositive, but in the developed world only 60–90% of adults show serological evidence of exposure to CMV.

Pathogenesis
Spread is by contact with infectious secretions (e.g. oropharyngeal) and through intrauterine and perinatal transmission (vaginal secretions and breast milk); the virus may also be acquired by blood transfusion or organ transplantation. The incubation period is 4–8 weeks. The primary infection is often asymptomatic. CMV remains latent in various tissues and may cause secondary infections, particularly in immunocompromised individuals.

Associated diseases
Infectious mononucleosis-like syndrome. Disseminated infection occurs in immunocompromised individuals (e.g. transplant recipients, acquired immune deficiency syndrome [AIDS] patients), leading to pneumonitis, hepatitis and retinitis. Congenital infections occur.

Diagnosis
This is made by: isolation in tissue culture; serology (IgM for active disease, IgG for immune status); tissue biopsy (typical 'owl's eye' inclusion bodies); nucleic acid detection methods (e.g. polymerase chain reaction [PCR]).

Treatment
Ganciclovir, foscarnet and cidofovir. CMV hyperimmunoglobulin (Chapter 31).

Epstein–Barr virus (EBV)

Virus
Enveloped DNA virus, 150–200 nm in diameter.

Epidemiology
EBV has a worldwide distribution. Seropositivity increases with age reaching > 90% in adults. Two peaks of seroconversion occur: 1–6 and 14–20 years.

Pathogenesis

Spread is by contact with saliva; the primary site of infection is the epithelium of the pharynx. B lymphocytes become infected, and EBV may lie dormant in these cells. The incubation period is 30–50 days. Atypical monocytes in the blood ('infectious mononucleosis') are mainly EBV-specific cytotoxic T lymphocytes.

Associated diseases

Glandular fever syndrome ('infectious mononucleosis'). EBV is implicated in the pathogenesis of several tumours including Burkitt's lymphoma and nasopharyngeal carcinoma, post-transplant lymphoproliferative disease and Hodgkin's disease. Glandular fever may be complicated by widespread EBV infection in a few patients (e.g. hepatitis, myocarditis, meningoencephalitis).

Diagnosis

Detection of mononuclear cells in peripheral blood; serology—test for heterophile antibodies (Paul–Bunnell test; antibodies that appear early in EBV infection and cross-react with sheep or horse erythrocytes, resulting in agglutination); immunofluorescence technique (IFT) or specific enzyme immunoassay (EIA) for antibody (IgG and IgM) to EBV capsid and nuclear antigens; nuclear acid detection methods.

Treatment

No effective antiviral treatment available.

Human herpesvirus (HHV) 6 and 7

Viruses

Enveloped DNA viruses, 150-200 nm. The two viruses show 90% sequence homology.

Epidemiology

Worldwide distribution. Infection early in childhood; 90% of children are seropositive for HHV6 by 2 years of age and for HHV7 by 5 years of age.

Pathogenesis

Infection through direct contact with saliva, and via blood products and organ transplantation. Viruses infect mainly T lymphocytes where they establish latent infection. Other cell targets include macrophages, natural killer cells and astrocytes.

Associated disease

Roseola infantum (exanthem subitum) in infants and young children. Meningitis, encephalitis, hepatitis and pneumonitis in immunocompromised patients particularly bone marrow transplant recipients.

Diagnosis

Isolation in tissue culture; serology, detection of IgG and IgM antibodies by IFA and EIA; nucleic acid detection.

Treatment

Symptomatic in children. Ganciclovir, foscarnet and cidofovir for complications in immunocompromised patients.

Human herpesvirus 8

Virus

Enveloped DNA virus 150–200 nm.

Epidemiology

Endemic in Africa and some parts of Italy, Greece, Spain and Brazil, with a seroprevalence of > 40% in adults. High seroprevalence among homosexual men in the developed world.

Pathogenesis

Transmission, in endemic areas, through close contact with saliva of infected people. In non-endemic areas sexual transmission is important, and possibly also via blood products and organ transplantation. Virus infects B lymphocytes and endothelial cells and establishes latent infection.

Associated disease

Kaposi's sarcoma, rare forms of B-cell lymphomas (body cavity-associated lymphoma, Castleman's disease).

Diagnosis
Serology, detection of latent and lytic antibodies by IFT and EIA; nucleic acid detection.

Treatment
No established antiviral treatment.

Influenza viruses

Virus
Enveloped RNA virus of 80–120 nm diameter; segmented genome (eight segments); there are three types, A, B and C, determined by nucleoprotein; type A infects humans and animals whereas types B and C infect humans only. The envelope contains two glycoproteins in lipid bilayer: haemagglutinin (HA) which attaches the virus to cellular receptors and neuraminidase (NA) which mediates the release of newly formed viruses from infected cells. Variations in HA and NA determine subtypes (in humans: H1, H2, H3, N1, N2). Nomenclature is based on type, origin, strain, year of isolation and subtype e.g. A/PR/8/34/(H1N1)

Type	Host	Origin	Strain no.	Year	Subtype
A	Human	Puerto Rico	8	1934	H1N1

Epidemiology
Antigenic variation is an important factor in global epidemiology.
• Antigenic 'drift': minor antigenic changes occur in HA and NA (within subtype) as a result of sequential point mutations (types A, B and C), leading to epidemics.
• Antigenic 'shift': major antigenic changes occur in HA and NA (change in subtype) as a result of genetic reassortment with animal viruses (type A only), leading to pandemics.

Pathogenesis and immune response
Spread by droplets; infection of upper respiratory tract; incubation period of 1–4 days; antibody in serum (IgM) and secretions (IgA) appears about day 6. HA is an important virulence determinant.

Associated disease
Influenza. Type C causes mild respiratory illness. Complications: pneumonia, primary viral pneumonia or secondary bacterial pneumonia, otitis media, myocarditis, Reye's syndrome, encephalitis.

Diagnosis
This is by virus isolation in tissue culture or embryonated eggs; antigen detection by immunofluorescence or EIA; serology (complement fixation test, haemagglutination inhibition test); nucleic acid detection.

Treatment
For patients at risk of severe influenza infection (elderly and immunocompromised individuals), amantadine, oseltamivir or zanamivir (Chapter 31).

Prevention
Inactivated vaccines (whole virus or subunit given intramuscularly) have an efficacy of 60–80% provided that vaccine components and current wild-type viruses are sufficiently similar. Also oseltamivir or zanamivir.

Paramyxoviruses

Virus
Spherical or pleomorphic enveloped RNA viruses, 150–300 nm in diameter, non-segmented genome

Classification
There are four genera:
1 Pneumovirus (respiratory syncytial virus [RSV])
2 Paramyxoviruses (parainfluenzaviruses types 1–4)
3 Rubulavirus (mumps virus)
4 Morbillivirus (measles virus).

Respiratory syncytial virus (RSV)

Virus
Two major strain groups A and B

Epidemiology
Worldwide distribution; seasonal activity during

winter months in temperate climates and throughout the year in warmer climates. Infects > 50% of infants aged < 1 year and by 3 years of age all children have experienced an RSV infection.

Pathogenesis
Transmission by large particle aerosols or fomites; incubation period 2–8 days; infects epithelium of upper respiratory tract and then spreads to lower respiratory tract. IgA, IgG and IgM responses not protective against recurrent infections and may play a role in the pathogenesis of disease.

Associated diseases
Bronchiolitis. Complications: pneumonia, exacerbation of asthma, severe disease in pre-term infants, infants with underlying pulmonary or cardiac disease, and immunocompromised patients.

Diagnosis
Isolation in tissue culture, antigen detection by immunofluorescence or EIA, serology (complement fixation test), nucleic acid detection.

Treatment
Generally symptomatic. Ribavirin for infants at risk of severe RSV infection (Chapter 31).

Prevention
Palivizumab (humanized mouse IgG monoclonal antibody) for children at risk of severe RSV infection before start of seasonal activity.

Parainfluenza viruses
Epidemiology
Type 3 annual epidemics during spring and summer, type 1 bi-annual outbreaks during winter. Fifty per cent of infants are infected by 1 year of age.

Pathogenesis
Transmission by aerosols; incubation period 2–8 days. Replication in nasopharyngeal epithelium followed by spread to tracheobronchial tree. Humoral immune response not protective and recurrent infections are common.

Associated diseases
Croup, bronchiolitis. Complications: pneumonia, otitis media, severe disease in immunocompromised children and adults.

Diagnosis
Isolation in cell culture, antigen detection by immunofluorescence, serology (haemagglutination inhibition), nucleic acid detection.

Treatment
Symptomatic

Prevention
No vaccine available

Measles
Virus
Single serotype.

Epidemiology
Worldwide distribution. Infections depend on the state of immunity and size of the population. Rapid epidemics occur when the virus is introduced into isolated, susceptible communities.

Pathogenesis
Transmission by large droplets; incubation period 10–12 days to prodromal symptoms and 10–21 days to appearance of rash. Infection starts in epithelium of upper respiratory tract followed by replication in local draining lymph nodes, and spreads to viscera and skin via the blood. IgA, IgG and IgM responses lead to life-long immunity.

Associated diseases
Complications of measles: pneumonia, encephalitis, subacute sclerosing panencephalitis (SSPE). Severity of illness increases in adults and immunocompromised patients.

Diagnosis
Isolation in tissue culture; serology (detection of IgM by EIA); nucleic acid detection.

Treatment
Symptomatic.

Prevention

Live attenuated virus vaccine (part of MMR) is used in mass immunization programmes. The vaccine is > 95% effective. The first immunization is given at the age of 15 months (to avoid failure as the result of the presence of maternal measles antibody); earlier application is given in developing countries, because there is a high measles mortality rate in children < 1 year old.

Mumps
Virus

Single serotype.

Epidemiology

Worldwide distribution. High incidence in children between 4 and 10 years of age; peak activity between January and May in temperate climates.

Pathogenesis

Transmission by large droplets; incubation period 16–18 days. Infection starts in epithelium of upper respiratory tract followed by replication in local draining lymph nodes, and spread to viscera by blood. IgA, IgG and IgM response lead to life-long immunity.

Associated diseases

Parotitis. Complications: meningitis, encephalitis, deafness, orchitis, oophoritis, pancreatitis.

Diagnosis

Isolation in tissue culture; serology (detection of IgM antibodies); nucleic acid detection.

Treatment

Symptomatic

Prevention

Live attenuated virus vaccine is available as the monovalent form (mumps only), or in combination with the rubella and measles vaccine (MMR).

Rhinoviruses

Viruses

Members of the Picornaviridae family of viruses.

Icosahedral, non-enveloped RNA viruses, 28–30 nm in diameter. There are more than 100 serotypes.

Epidemiology

Rhinoviruses are endemic worldwide. In temperate climates two peaks of infection are documented: in the autumn and in early spring.

Pathogenesis

Transmission is by droplets or close contact. The incubation period is 2–4 days. Replication takes place in the nasal mucosa with the release of several proinflammtory cytokines, leading to characteristic symptoms of the common cold.

Associated disease

Causes 30–60% of common colds. Complications: acute otitis media, severe respiratory illness in infants with bronchopulmonary dysplasia and immunocompromised patients, exacerbation of asthma.

Diagnosis

Isolated in cell culture; nucleic acid detection.

Treatment

Symptomatic.

Coronaviruses

Virus

Enveloped RNA viruses 120 nm in diameter. Surface projections look like a crown, hence the name— coronaviruses. Three major strains: human coronaviruses (HCoV) 229E and OC43, and severe acute respiratory syndrome coronavirus (SARS-CoV).

Epidemiology

HCoV has a worldwide distribution and infections occur throughout the year, affecting all age groups equally. An SARS-CoV epidemic started in 2002 in southern China from where it spread to Hong Kong, Vietnam, Singapore, Canada, Taiwan, Thailand and other parts of the world. The epidemic ended in 2003 affecting 8500 patients (95% in Asia) with a mortality rate of 9.5%.

Pathogenesis

The incubation period is 2–5 days for HCoV and 2–10 days for SARS-CoV. Transmission by aerosols for HCoV and by close contact, droplets and possible faecal–oral route for SARS-CoV. Replication takes place in the epithelial cells of the respiratory tract and gut with development of strain-specific antibody. Asymptomatic shedding in faeces is common.

Associated infection

Common cold. HCoV is responsible for 25% of cases of colds. SRAS-CoV causes a syndrome characterized by fever, cough, dyspnoea, hypoxaemia, pneumonia and diarrhoea.

Diagnosis

This is determined by: growth in tissue culture; electron microscopy; serology (IFA, EIA and immunoblot); nucleic acid detection.

Treatment

Symptomatic

Control

No specific vaccine.

Adenoviruses

Virus

Non-enveloped DNA virus, 70–100 nm in diameter. There are 51 human adenovirus serotypes.

Epidemiology

Adenoviruses are endemic worldwide. Infection is spread via the faecal–oral route or droplets. Epidemics have been observed among closed communities such as boarding schools and military recruits. Frequent infection occurs in transplant recipients.

Pathogenesis

The incubation period is 2–15 days. Infection is of the epithelial cells of the respiratory and gastrointestinal tract.

Associated diseases

Pharyngitis; pharyngoconjunctival fever and keratoconjunctivitis; gastroenteritis (types 40 and 41); pneumonia, hepatitis and haemorrhagic cystitis, especially in transplant recipients.

Diagnosis

This is made by: virus isolation from throat swabs or faeces in cell culture; direct detection of viral antigen in respiratory secretions by immunofluorescence or in stool samples by EIA; detection by electron microscopy (faeces and urine); detection of viral antibody by serology (e.g. complement fixation test); nucleic acid detection methods.

Treatment

Symptomatic.

Enteroviruses

This genus of viruses includes:
- polioviruses (types 1, 2, 3)
- coxsackie A viruses, 23 types (types 1–22, 24)
- coxsackie B viruses, 6 types (types 1–6)
- echoviruses, 31 types (types 1–9, 11-27, 29-33)
- enteroviruses, 4 types (types 68–71).

Viruses

Members of the Picornaviridae family of viruses. Icosahedral, non-enveloped RNA viruses, 28–30 nm in diameter.

Epidemiology

Enteroviruses have worldwide distribution. Infections occur in infancy in developing countries and in early childhood in developed countries.

Pathogenesis

The incubation period is from 2 to 40 days; normal habitat and primary site of replication are the intestinal tract; primary replication takes place in the oropharynx and intestine, followed by viraemia and infection of target organs, i.e. the central nervous system (CNS: brain, meninges and spinal cord) or muscle tissue (heart, skeletal muscle) and skin. Long-lasting local IgA and humoral IgM/IgG response develops after natural infection.

Associated diseases
Poliomyelitis, aseptic meningitis, encephalitis, myocarditis, neonatal sepsis, pleurodynia, herpangina, hand, foot and mouth disease, and conjunctivitis.

Diagnosis
Isolation in cell culture; serology by enterovirus IgM-specific EIA or neutralization tests; nucleic acid detection.

Prevention and control
Both live (Sabin) and killed (Salk) vaccines are available and successfully used by the World Health Organization (WHO) in its poliomyelitis eradication programme.

Treatment
Pleconaril in cases with severe complications, i.e. meningitis (Chapter 31).

Rubella virus

Virus
Enveloped RNA virus, diameter 60 nm. Single serotype.

Epidemiology
Worldwide distribution. A frequent childhood infection particularly among schoolchildren. Incidence in the developed world markedly reduced since immunization programmes started, but in the developing world 25% of women at childbearing age are susceptible to rubella.

Pathogenesis
Transmission by large droplets. The incubation period is 13–20 days, with primary replication in the epithelium of upper respiratory tract and the cervical lymph nodes; the rash coincides with the appearance of serum antibody. Lifelong immunity follows natural infection. Primary infection during the first 12 weeks of pregnancy often leads to generalized fetal infection, leading in turn to cataract, nerve deafness, cardiac abnormalities, hepatosplenomegaly, purpura or jaundice (congenital rubella syndrome).

Associated disease
Rubella: rash and lymphadenopathy with transient arthralgia. Complications: congenital rubella syndrome, post-infectious encephalopathy.

Diagnosis
• Acute infection: specific IgM antibody is determined by EIA; virus isolation in tissue culture; nucleic acid detection.
• The immune status is determined by detection of specific IgG antibody by EIA.

Treatment
Symptomatic. Proven infection during the first 3 months of pregnancy is an indication for therapeutic abortion.

Prevention and control
Live attenuated vaccine is used. Immunization programmes exist in most developed countries.

Parvoviruses

Virus
Small, round, non-enveloped DNA viruses of diameter 18–25 nm; single serotype: human parvovirus B19.

Epidemiology
Worldwide distribution. Infection common in children aged between 4 and 10 years. About 40% of children are seropositive by the age of 15 years. In elderly people, > 90% are seropositive

Pathogenesis
The incubation period is up to 17 days. Transmission is by respiratory secretions and blood products. Replicates in rapidly dividing cells, mainly erythroid progenitor cells.

Associated diseases
Erythema infectiosum or fifth disease: fever, chills, myalgia 1 week after infection, followed by

maculopapular rash at 17 days. Arthropathy in adults, particularly women. Aplastic crisis can occur in patients with chronic haemolytic anaemias (e.g. thalassaemia, spherocytosis). Hydrops fetalis, spontaneous abortions and intrauterine death may result from fetal infection. Chronic infection with pure red cell aplasia in immunocompromised patients.

Diagnosis
Serology; detection of specific IgG and IgM antibodies by EIA; nucleic acid detection.

Treatment
Human immunoglobulins in cases of persistent infection in immunocompromised patients.

Rotaviruses

Virus
RNA virus, 75 nm in diameter with wheel-like structure. Segmented genome (11 segments). Five groups (A–E), and within group A at least two subgroups (I, II) and various serotypes (including G and P types; G1–G4 viruses cause over 90% of human infections).

Epidemiology
Rotaviruses occur worldwide with different serotypes co-circulating, usually affecting children < 2 years with a winter peak in temperate climates.

Pathogenesis and immune response
Infection is via the faecal–oral route with an incubation period of 1–2 days. Virus replication occurs in the epithelium of the small intestine and results in the release of large numbers of particles in human faeces (10^{11} particles/gram). There are local (IgA) and humoral (IgM/IgG) serotype-specific and cross-reactive immune responses.

Associated disease
Gastroenteritis: self-limiting diarrhoea and vomiting lasting 4–7 days; mild-to-severe dehydration can occur (depending on strain).

Diagnosis
This is made by direct detection in faeces by electron microscopy, ELISA, latex agglutination test and nucleic acid detection methods.

Treatment
Rehydration with oral rehydration fluid.

Prevention and control
Increased personal and water hygiene helps control outbreaks. Vaccines are under development.

Noroviruses and saproviruses

Viruses
Small non-enveloped RNA viruses, diameter 27–35 nm. Members of the caliciviridae family of viruses characterized by cup-shaped depression on surface of particle.

Epidemiology
Worldwide distribution. Outbreaks occur year round but more frequently during winter months in temperate climates. Outbreaks associated with contaminated water supplies and food. More than 30% of children are seropositive by 3 years of age, > 80% are seropositive in early adulthood. Most common cause of gastroenteritis outbreaks in hospitals.

Pathogenesis
Transmission by faecal–oral route and incubation period is 16–48 h. Viral replication in epithelial cells of the villi of the small intestine.

Associated diseases
Gastroenteritis characterized by projectile vomiting and diarrhoea lasting for 24–48 h.

Diagnosis
Detection in stool by electron microscopy, EIA and nucleic acid detection methods.

Treatment
Symptomatic.

Prevention
No vaccine available

Hepatitis viruses

Hepatitis A virus (HAV)

Virus
Member of picornavirdae family of viruses, genus *Hepatovirus*. Non-enveloped RNA virus 28–32 nm (previously enterovirus 72); only one serotype.

Epidemiology
Most infections in endemic areas (developing world) occur before 5 years of age and the majority are asymptomatic. In the developed world most clinical cases occur in adults. Common source outbreaks result from contamination of drinking water and food. Infections acquired frequently by travellers from non-endemic to endemic areas. Antibody prevalence in young adults is 30–60%, higher in lower socioeconomic groups.

Pathogenesis
Faecal–oral transmission occurs with an incubation period of 3–5 weeks (mean 30 days). The virus is present in blood from 2 weeks before to 1 week after jaundice, and slightly longer in faeces. All age groups are susceptible to hepatitis A infection and disease severity increases with age. In most cases there is complete recovery, with specific antibody response persisting lifelong. There is no chronic disease or carrier state.

Associated disease
Acute hepatitis.

Diagnosis
This is made by serology (EIA) testing for specific HAV IgM (acute infection) or IgG (immune status).

Treatment
Symptomatic.

Prevention
A killed HAV vaccine is available.

Hepatitis B virus (HBV)

Virus
Member of hepadnaviridae family of viruses. DNA virus, , 42 nm in diameter, consisting of:
- core, DNA circular genome
- nucleocapsid (hepatitis B core antigen, HBcAg); 'e' antigen (HBeAg) is a cleavage product of the core antigen found on infected cells or free in serum
- envelope (hepatitis B surface antigen, HBsAg).
- also exists as 22 nm spherical or filamentous particle consisting of hepatitis B surface antigen, HBsAg.

Epidemiology
Hepatitis B virus has worldwide distribution, with more than 200 million carriers (prevalence in north and mid-Europe and North America 0.1–0.5%; southern Europe 2–5%; Africa and south-east Asia 6–20%).

Pathogenesis
- Transmission is parenteral (via blood and blood products), by sexual intercourse and vertical through passage in the birth canal (this is the main route of transmission in Asia and Africa).
- Incubation period of 50–180 days.
- Viral replication in liver results in lysis of hepatocytes by cytotoxic T cells.
- Hepatic damage reversed 8–12 weeks in ≥ 90% cases; 2–10% become chronic carriers (persistence of HBsAg for > 6 months).
- 95% of newborns of carrier mothers become carriers if untreated.

Associated diseases
Acute, chronic and fulminant hepatitis; hepatic cirrhosis; and hepatocellular carcinoma.

Diagnosis
Serological tests are made by immunoassays detecting HBsAg, HBeAg, and antibodies to HBcAg (IgM and IgG), anti-HBeAg and anti-HBsAg; nucleic acid detection.

Prevention
HBV vaccines. HBV immunoglobulin for post-exposure prophylaxis and neonates of carrier mothers

Treatment
Interferon-α, lamivudine or adefovir (Chapter 31).

Hepatitis C virus (HCV)

Virus
RNA virus, 4–50 nm in diameter; a new genus of the Flaviviridae family of viruses. High genome variability (at least six different genotypes 1–6 and several subtypes 1a, 1b, 2a, 2b, etc.).

Epidemiology
HCV occurs worldwide. Antibody prevalence varies between < 1% in the USA and western Europe and 2% in southern Italy, Spain and central Europe. Higher prevalence rates, up to 20%, are detected in Egypt. Also high rates of infection among intravenous drug users in most parts of the world.

Pathogenesis and associated diseases
HCV has a similar pathogenesis to HBV; but, in contrast to HBV, infections are followed by chronic hepatitis in 60–80% of cases.

Diagnosis
This is made by serology, i.e. EIA to detect HCV antibodies, and by nucleic acid detection methods.

Prevention
Since 1991, routine testing of all blood and organ donors for HCV antibody has been undertaken to prevent transmission by transfusion or transplantation.

Treatment
Interferon-α and ribavirin.

Delta agent ('hepatitis D virus', HDV)

Virus
Defective RNA virus, 36 nm in diameter, which replicates only in HBV-infected cells.

Epidemiology
Worldwide distribution. High prevalence in the Mediterranean area, Africa, South America, Japan and the Middle East. It has similar transmission and at-risk groups to HBV.

Pathogenesis and associated disease
Infections is either a co-infection with HBV or a superinfection of chronic HBV infection leading to aggravation of HBV disease.

Diagnosis
Serology (EIA) can be used to detect HDV antibody and HDV antigen, nucleic acid detection methods.

Prevention
HBV vaccine

Hepatitis E virus (HEV)

Virus
Morphology, size and genome organization resemble caliciviruses but still remain unclassified. Single serotype.

Epidemiology
Endemic in Indian subcontinent, south-east Asia, Middle East, north Africa and central America. Common source outbreaks caused by contaminated water or food are common. In developed countries, sporadic cases are detected among travellers returning from endemic areas.

Pathogenesis
Transmission by faecal–oral route and rarely by blood transfusion in endemic countries. Incubation period 6 weeks. Specific IgG and IgM antibodies are produced.

Associated diseases
Acute self-limiting hepatitis with no evidence of chronic infection. High mortality rate (10–20%) in pregnant women.

Diagnosis
Serology (detection of IgG and IgM antibodies by EIA); nucleic acid detection.

Treatment
Symptomatic.

Prevention
Vaccine under development.

Human retroviruses

Human immunodeficiency virus (HIV)
(See also Chapter 47)

Virus
A member of the Lentivirinae subfamily of the Retroviridae. Enveloped RNA viruses 100–150 nm in diameter. Replicates through DNA intermediates utilizing the RNA-directed DNA polymerase (reverse transcriptase). There are two types: HIV-1 and HIV-2 (40% sequence homology). HIV-1 is divided into three groups: M group (11 subtypes A–K), and O and N groups. Group M subtypes have worldwide distribution whereas groups O and N are confined to parts of west and central Africa. Generally, HIV isolates are obtained from different people, sequentially from the same person, and even within individual isolates have a high degree of genomic variability (quasi-species).

Epidemiology
There has been worldwide spread of HIV since 1981. Estimates of prevalence worldwide vary greatly. By the end of 2005, 40 million people worldwide were estimated to be living with HIV (25 million in Africa and 7.5 million in Asia); 20 million people have died of HIV since the epidemic started in 1981.

The main risk groups for HIV infection are:
- homosexuals
- intravenous drug abusers
- people with haemophilia and blood transfusion recipients (before 1985)
- sexually promiscuous individuals
- children born to HIV-infected mothers
- heterosexual contacts of HIV-infected individuals.

Pathogenesis and immune response
Transmission is mainly by infected blood or blood products, through sexual intercourse or from mother to child during pregnancy. HIV chronically infects CD4+ cells (T4 lymphocytes, helper and inducer cells, monocytes, macrophages, dendritic cells and microglial cells) by binding to CD4 molecules as a receptor. This results in early impairment of various T4-cell functions followed, by a decrease in T4-cell numbers. A humoral antibody response against most HIV-specific proteins develops. In later stages deficiency of both the humoral and the cellular immune responses leads to the development of AIDS.

Associated diseases
AIDS characterized by severe disease resulting from generalized infections with bacteria (mycobacteria), viruses (HSV, HVC, VZV), fungi (*Candida*, *Aspergillus*, *Cryptococcus*), protozoa (*Pneumocystis*, *Toxoplasma*) and associated with tumours (Kaposi's sarcoma, lymphomas).

Diagnosis
Detection of HIV-specific antibody and antigen by passive particle agglutination tests (PPAT), EIA, and western blotting (WB). Molecular techniques for detecting HIV, quantifying viral load and typing for drug resistance have become a significant part of patient diagnosis and clinical management.

Treatment
Four classes of drugs are available: nucleoside reverse transcriptase inhibitors (e.g. zidovudine, lamivudine), non-nucleoside reverse transcriptase inhibitors (e.g. nevarapine and delavirdine), protease inhibitors (e.g. indinavir, retonavir) and fusion inhibitors (T20-enfuvirtide). A combination of three drugs (highly active antiretroviral therapy or HAART) is the accepted standard form of treatment.

Prevention and control
- There is no antiviral cure.
- No effective vaccine has been developed; there are problems as a result of viral variability, latency and evasion to immune response.
- Testing of all blood and organ donors prevents transmission from those sources.

- Information campaigns, needle exchange programmes and condom use.
- Post-exposure prophylaxis with HAART is recommended after exposure in healthcare settings.

Human T-cell lymphotropic viruses HTLV-I and HTLV-II

Virus

Members of the Oncovirinae subfamily of the Retroviridae. RNA genome less variable than that of HIV; HTLV-II shows 60–70% sequence homology with HTLV-I.

Epidemiology

HTLV-I is endemic in Japan, the Caribbean, Melanesia, part of sub-Saharan Africa and Brazil. HTLV-II is endemic among native American Indians in North, Central and South America. High seroprevalence is also detected among intravenous drug users in North America.

Pathogenesis

Transmission by blood (cell containing blood and not plasma), sexual intercourse and from mother to child through breast-feeding. HTLV-I can immortalize T lymphocytes in vitro and can transactivate T cells in vivo, but its genome does not contain oncogenes. No similar characteristics have been proved for HTLV-II. Most infected individuals remain asymptomatic carriers for life. Only 2–4% of infected people develop adult T-cell leukaemia and lymphoma (ATLL) after many decades of infection. HTLV-I can invade the CNS and the resulting inflammatory immune response is responsible for the pathogenesis of HTLV-associated myopathy (HAM).

Associated diseases

ATLL, HAM, uveitis. HTLV-II has been associated with some cases of HAM but the association is not definite.

Diagnosis

Detection of HTLV-I-specific antibody is made by EIA, particle agglutination assays and western blotting; nucleic acid detection methods.

Prevention and control

Testing of all blood donors is undertaken in some countries (i.e. the USA and the UK).

Treatment

No antiviral drugs available.

Human papillomavirus (HPV)

Virus

Non-enveloped DNA virus, 55 nm in diameter. Over 100 genotypes.

Epidemiology

HPV occurs worldwide.

Pathogenesis

Transmission is by close direct contact, sexual intercourse and perinatal transmission by passage through the birth canal. Infects and replicates in squamous epithelium on both keratinized and mucosal surfaces. Persistent and latent infection and malignant transformation of infected cells.

Associated disease

Skin warts (e.g. plantar warts), and other skin conditions (epidermodysplasia verruciformis), genital warts (e.g. condyloma acuminata), and cervical, penile and anal carcinoma. Laryngeal papillomatosis.

Diagnosis

Cytology and colposcopy; serology; nucleic acid detection.

Therapy

Surgery, cauterization, cryotherapy and laser therapy; interferon, imiquimod (podophyllum resin).

Prevention

Vaccines are under development.

Rabies virus

Virus

Member of the Rhabdoviridae family. It is a bullet-shaped RNA virus (75 × 180 nm). Several genotypes (include classic rabies, European, Australian bat rabies and rare African genotypes). All known to cause disease in humans but the most common source is classic rabies.

Epidemiology
Rabies virus has a wide animal reservoir (e.g. foxes, skunks, racoons, bats) and is enzootic in terrestrial mammal species worldwide. Some parts of the world are considered free from rabies, i.e. Ireland, New Zealand. The UK used to be free from rabies but recently three isolations of bat rabies were made, including a fatal case in a Scottish bat conservationist. Human infections occur mostly in Asia, Africa and South America, and dogs are the cause of > 90% of these infections.

Pathogenesis
Infection mainly occurs through the bite of a rabid animal, and rarely after corneal transplant from infected patients. Incubation period can be as short as 4 days to 19 years, but is generally 20–90 days. Multiplication takes place in muscle cells and at neuromuscular junctions. The virus migrates along peripheral nerves to the CNS. Replication in the CNS (mainly basal ganglia, pons, medulla) causes nerve cell destruction and inclusion bodies (Negri bodies). Finally migration from the CNS along peripheral nerves to different tissues, including peripheral nerves, skeletal and cardiac muscles and salivary glands, occurs.

Associated disease
Paraesthesiae; anxiety; hydrophobia; paralysis; coma. Mortality rate of virtually 100%.

Diagnosis
Virus isolation, detection of viral antigens by immunofluorescence, serology (rabies antibodies in blood and CSF), nucleic acid detection.

Therapy and control
• Pre-exposure prophylaxis: immunization of at-risk groups (veterinary personnel, laboratory workers, travellers to endemic areas)—three doses (0, 7, 28 days) of rabies vaccine followed by a booster dose 1 year later.
• Post-exposure prophylaxis: five doses of vaccine (0, 3, 7, 14, 28 days) plus one dose of rabies immunoglobulin.
• Spread can be controlled by quarantine of animals (the UK) and immunization of domestic animals (dogs, cats) in endemic areas.

Chapter 20

Basic mycology and classification of fungi

Characteristics

Fungi are eukaryotic organisms; they are distinct from plants in not containing chlorophyll. Fungi are macroscopic (mushrooms) or microscopic (moulds and yeasts). Only a few species cause human disease. Fungi are non-motile; they may grow as single cells (yeasts) or filamentous structures (mycelia), some of which may be branched.

Classification

Fungi are classified according to their method of sexual reproduction. Those that do not reproduce sexually are called 'fungi imperfecti'; those that reproduce sexually may be self-fertile or require strains of an opposite type to allow sexual fusion to occur. These are four groups of fungi that cause human diseases (Table 20.1).

Fungal infections

Fungal infections can be divided clinically into two groups: superficial and deep mycoses (Tables 20.2 and 20.3). Most deep mycoses are opportunist infections occurring in immunocompromised patients.

Laboratory diagnosis
● *Direct microscopy* of clinical material, including skin scrapings and sputum; the characteristic morphology facilitates identification.
● *Culture*: fungi can be grown on most routine media (Plates 31 and 32), but may require a prolonged incubation. Antibiotic-containing selective media (e.g. Sabouraud's glucose agar), which inhibit bacterial growth, are often used. Incubation at both 37°C and 28°C facilitates the isolation of common filamentous fungi and yeasts.
● *Serology*: serological tests are available for the diagnosis of some fungal infections (e.g. candidiasis and aspergillosis), but these lack specificity and sensitivity.

Examples of fungal pathogens

Cryptococcus neoformans

● *C. neoformans* is a capsulate yeast (Plate 30) found worldwide. Its natural habitat is soil, particularly soil contaminated with bird droppings.
● *C. neoformans* has a polysaccharide capsule that can be visualized by mixing the organism in Indian ink. It can be observed in cerebrospinal fluid (CSF) by this method; it can be isolated from sputum, bronchoalveolar lavage fluid, tissue biopsies and other methods of detection, including latex agglutination of CSF or serum.
● Infection is acquired by inhalation of the fungus. Human infections are rare; most occur in immunocompromised individuals, including those with HIV.

Table 20.1 Four groups of fungi that cause human disease.

FUNGI		
Group	**Comment**	**Examples**
Yeasts	Have round or oval cells that multiply by budding	*Cryptococcus neoformans* (Plate 33)
Yeast-like fungi	Grow predominantly as yeasts that bud; may also form chains of elongated filamentous cells called pseudohyphae	*Candida albicans* (Plates 29 & 33)
Filamentous fungi	Grow as filaments (hyphae) and produce an intertwined network called a mycelium; produce asexual spores (conidia), which may be single or multi-celled. Conidia are produced in long chains on aerial hyphae (conidiophores)	*Aspergillus, Trichophyton* and *Zygomycetes* (including *Mucor*)
Dimorphic fungi	Have two forms of growth: filamentous at 22°C (saprophytic phase) and yeast like at 37°C (parasitic phase)	*Blastomyces, Coccidioides, Histoplasma capsulatum*

Table 20.2 Superficial mycoses.

SUPERFICIAL MYCOSES			
Fungi	**Type of fungus**	**Principal infections**	**Epidemiology**
Candida albicans	Oral thrush, vaginitis, cutaneous candidiasis	Worldwide	
Dermatophytes *Epidermophyton Microsporum Trichophyton*	Filamentous	Tinea (ringworm) of skin and hair	Worldwide
Malassezia furfur	Dimorphic	Pityriasis versicolor	Worldwide, most common in the tropics

Table 20.3 Deep mycoses.

DEEP MYCOSES			
Fungi	**Type of fungus**	**Principal infections**	**Epidemiology**
Aspergillus fumigatus	Filamentous	Lung infection, disseminated aspergillosis	Worldwide
Candida albicans (Plate 29)	Yeast-like	Lung infection, oesophagitis, endocarditis, candidaemia with disseminated candidiasis	Worldwide
Cryptococcus neoformans (Plate 30)	True yeast	Meningitis	Worldwide
Histoplasma capsulatum	Dimorphic	Lung infection	USA
Coccidioides immitis	Dimorphic	Lung infection	Central/South America
Blastomyces dermatidis	Dimorphic	Lung infection	Africa, America
Paracoccidioides brasiliensis	Dimorphic	Lung infection	South America
Rhizopus arrhizus	Filamentous and related fungi	Rhinocerebral infection, lung infection, disseminated infection (mucormycosis)	Worldwide

- Primary infection occurs in the lungs and is usually asymptomatic. Acute pneumonia may occur with fungaemia and infection in various organs, particularly the brain and meninges. Treatment is with amphotericin.

Candida

- The genus *Candida* contains a number of species, including *C. albicans* (the most frequently isolated pathogen), *C. parapsilosis* and *C. tropicalis*. *C. dubliniensis* has emerged as a pathogen in patients with the acquired immune deficiency syndrome (AIDS). *C. albicans* (Plate 29) is a commensal of the mouth and gastrointestinal tract.
- Superficial candida infections are common and include vaginal and oral candidiasis (thrush), skin and nail infections, when these infections may arise because the hands are in the water for long periods, and a complication of antibiotic therapy that temporarily reduces the bacterial flora.
- Invasive candida infections may involve the gastrointestinal tract (e.g. oesophagus), lungs and urinary tract. Candidaemia may result in abscesses in various organs (e.g. brain, liver). These infections occur primarily in immunocompromised patients.
- Candida can also colonize prosthetic materials, e.g. intravascular catheters and peritoneal dialysis cannulae, resulting in septicaemia and peritonitis, respectively. *Candida* is a rare cause of endocarditis.

Laboratory diagnosis
By direct microscopy of appropriate clinical material for oval Gram-positive cells, some of which may be budding or producing pseudomycelia; culture (Plate 33); and serology for candida antibodies in patients with deep-seated infections. Pseudomycelia may also be seen with appropriate stains in histopathological specimens.

Treatment
Topical with nystatin; parenteral therapy is with fluconazole, itraconazole, voriconazole, caspofungin or amphotericin B.

Malassezia furfur

- *M. furfur* is part of the normal human flora. It produces thick-walled, budding cells and curved hyphae.
- It causes pityriasis versicolor, a superficial scaly skin infection with depigmentation.

Diagnosis
By microscopy of skin scales showing yeast cells.

Treatment
Topical or oral with azole antifungals.

Aspergillus

- The genus *Aspergillus* contains a number of species, including *A. fumigatus* (the most frequent human pathogen) and *A. niger* (see Plate 32).
- It is a common saprophyte worldwide, frequently found in soil and dust. Outbreaks of aspergillosis in immunocompromised patients have occurred because of construction work adjacent to hospitals, which can result in the release of large numbers of spores.
- *A. niger* is common in ear infections.
- *A. fumigatus* infections are acquired by inhalation of spores (conidia) (Plate 34), resulting in diffuse lung infection or, occasionally, a large mycelial mass (aspergilloma).
- Infection can also spread to other sites, including adjacent blood vessels and sinuses, or become disseminated to the liver, kidneys and brain (Plate 35). The fungus also causes chronic infections of the ear.
- *Aspergillus* infections are most frequent in immunosuppressed patients, e.g. patients with leukaemia, transplant recipients and patients with AIDS.
- *Aspergillus* is also associated with allergic alveolitis, which occurs in atopic patients with recurrent exposure to aspergillus spores; symptoms include fever, cough and bronchospasm.

Laboratory diagnosis
By: direct examination of appropriate samples for branching hyphae; culture; serology (of limited

value). *Aspergillus* may also be seen in histopathological specimens

Treatment
Amphotericin B, itraconazole or voriconazole.

Dermatophytes

The dermatophytes are a group of related filamentous fungi, also referred to as the ringworm fungi, which infect skin and related structures. Three clinically important genera have been described (Table 20.4) (Plates 36–38).

Epidemiology
The natural habitat is humans, animals or soil; human infection results from spread from any of these reservoirs. Dermatophyte infections are found worldwide with different species predominating in various climates.

Laboratory diagnosis
• Skin scrapings, or hair or nail clippings from active lesions are examined microscopically in 30% potassium hydroxide on a glass slide; the presence of hyphae confirms the diagnosis. Occasionally, the dermatophyte species can be identified by typical morphology (see Plate 38).
• Samples can be cultured on Sabouraud's medium at 28°C. Subsequent species identification is based on growth rate, colony appearance and microscopic morphology (see Plate 36).
• Infected hair may fluoresce under ultraviolet light (Wood's light) and is characteristic of certain infections, e.g. *M. canis*.

Treatment
Depends on the site and severity of infection. Options include topical imidazoles (e.g. clotrimazole, miconazole); oral griseofulvin, itraconazole or terbinafine.

Mucormycosis

The term 'mucormycosis' (or 'zygomycosis') refers to infections caused by a variety of filamentous fungi belonging to the order Mucorales. Medically important species include *Rhizopus arrhizus* and *Absidia corymbifera*.

Epidemiology
The Mucorales are found worldwide, in soil and decaying organic matter. Infections occur primarily in immunocompromised individuals. As with *Aspergillus*, hospital outbreaks of infection have occurred in association with building work.

Risk factors
Diabetes mellitus, burns, neutropenia.

Table 20.4 Clinically important genera of dermatophytes.

DERMATOPHYTES		
Genera	**Example**	**Infection**
Epidermophyton	*E. floccosum*	The only important species. It infects the skin (tinea corporis), nails (tinea unguium), groin (tinea cruris) and feet (tinea pedis) (Plate 36)
Microsporum	Infect hair and skin	*M. audouini* causes epidemic ringworm of the scalp (tinea capitis) in children. *M. canis*, which is a parasite of cats and dogs, occasionally causes ringworm in children
Trichophyton	Infect skin, hair and nails	*T. mentagrophytes* is the most common cause of tinea pedis (Plates 37 and 38). *T. rubrum* causes severe recurrent skin and nail infections
T. violaceum *T. tonsuras*		Common cause of tinea captis in Asian subcontinents, North America and the West Indies

Infection
- Pulmonary: an often fatal infection of immuno-compromised patients may result in disseminated infection (e.g. brain, liver, gastrointestinal tract).
- Rhinocerebral: an infection of the nasal sinuses may spread rapidly to involve the face, orbit and brain. It occurs particularly in uncontrolled diabetes mellitus and is often fatal if treatment is delayed.
- Necrotic tissue from burns (broad, non-septate filaments with right-angle branching).

Laboratory diagnosis
Microscopy and culture of appropriate specimens are undertaken (including necrotic lesions, sputum, bronchoalveolar lavage [BAL]).

Treatment
Amphotericin, plus urgent débridement of necrotic tissue in rhinocerebral infection.

Coccidioides immitis

C. immitis is a dimorphic fungus found in the soil in hot arid areas of south-west USA, and Central and South America.

Infection
This follows inhalation of arthrospores and is often subclinical. A mild, self-limiting pneumonia, often accompanied by a maculopapular rash, occurs in some patients. Severe progressive disseminated disease (e.g. meningitis, osteomyelitis) may occur, principally in immunosuppressed patients. Pulmonary infection may occur in AIDS patients who have lived or travelled through endemic regions.

Laboratory diagnosis
Direct microscopy and culture of appropriate specimens (culture should be attempted only in specialized centres because of the risk of infection); serology (e.g. latex agglutination) for IgM antibodies is available at specialist laboratories.

Treatment
Amphotericin, itraconazole and fluconazole.

Blastomyces dermatitidis

B. dermatitidis is a dimorphic fungus found in the soil in North and South America, and Africa. Humans are probably infected by inhalation.

Infections
Primary pulmonary infection may be complicated by haematogenous spread to involve the skin (granulomatous ulcers), bone and joints, brain and other organs.

Laboratory diagnosis
Microscopy and culture of appropriate specimens are undertaken. Serology is available at specialist laboratories but is of limited value.

Treatment
Amphotericin B or itraconazole.

Histoplasma capsulatum

H. capsulatum is a dimorphic fungus found worldwide, but infections occur most commonly in North, Central and South America. The natural habitat is soil, particularly in sites enriched with bat droppings (e.g. caves).

Infections
Infection is acquired by inhalation of microconidia, which germinate in the lung to produce budding yeast cells. Pulmonary infection is normally self-limiting, but chronic pulmonary disease with cavitations (similar to tuberculosis) may occur in patients with underlying lung disease. Disseminated histoplasmosis (liver, bone, brain, skin) may occur, particularly in immunocompromised patients.

Laboratory diagnosis
Microscopy of stained blood films or histological sections of tissue (oral yeast forms seen within mononuclear phagocytes); culture; serology (e.g. complement fixation test).

Treatment
Amphotericin or itraconazole.

Paracoccidioides brasiliensis

A dimorphic fungus found in the soil in Central and South America, *P. brasiliensis* causes pulmonary infection and mucocutaneous lesions including ulceration of the mucous membranes of the nasal and oral pharynx, which may progress to destruction of the palate and nasal septum. Disseminated infections occur with haematogenous spread to various sites, including the spleen, liver, bone and brain.

Laboratory diagnosis
Direct microscopy of pus, sputum or tissue biopsy; culture; serology (e.g. complement fixation test).

Treatment
Itraconazole, or amphotericin plus sulphadiazine.

Pneumocystis jiroveci (previously named *P. carinii*)

Classification
A unicellular fungus; forms trophozoites and cysts.

Epidemiology
Worldwide distribution; transmission is probably by droplet inhalation, but understanding of the epidemiology is incomplete. One theory suggests that *P. jiroveci* colonizes the lungs of many individuals and infection in immunocompromised individuals represents reactivation. Extrapulmonary infection at other sites occurs rarely.

Clinical conditions
Opportunist lung infection in immunocompromised individuals and severely malnourished children is found; extrapulmonary infections occur rarely (e.g. heart, liver, kidneys and eyes). In HIV infection risk of pneumonia increases if CD4+ cell count is < 200 mm^3.

Diagnosis
By direct examination of bronchial biopsies, washings or bronchoalveolar aspirates or open lung biopsies. Parasites are identified by induced sputum, histopathological stains or specific fluorescein-labelled antibodies.

Treatment
Co-trimoxazole.

Prevention
Required for certain immunocompromised patients (e.g. AIDS patients, after transplantation); use co-trimoxazole or nebulized pentamidine.

Chapter 21

Parasitology: protozoa

Classification

Parasites are classified into two subkingdoms (Table 21.1):

1 Protozoa: unicellular organisms.
2 Metazoa: multicellular organisms with organ systems.

Protozoa

Classification
Protozoa are classified into four groups according to their structure (Tables 21.1 and 21.2).

Amoebae

Classification
Many amoebae are human commensals (e.g. *Entamoeba coli, Endolimax nana*). *Entamoeba histolytica* is an important human pathogen. Some free-living amoebae, e.g. *Naegleria fowleri* and *Acanthamoeba* species (Plate 39), are opportunistic human pathogens .

Structure/physiology
Amoebae are unicellular microorganisms, with a simple two-stage life cycle:

1 Trophozoite: actively motile pleomorphic feeding stage.
2 Cyst: infective stage.
Division occurs by binary fission of the trophozoite or production of multiple trophozoites in a multinucleated cyst. Amoebae are motile via the formation of pseudopods. Cyst formation occurs under adverse conditions.

Entamoeba histolytica
Epidemiology
E. histolytica has worldwide distribution, primarily in subtropical and tropical regions. Infected cases (symptomatic or asymptomatic carriers) act as a reservoir. Spread is usually via water or food contaminated with cysts, and occasionally via oral–anal sex.

Pathogenesis
Cysts are ingested and gastric acid promotes release of the trophozoite in the small intestine. Trophozoites multiply and may cause necrosis and ulceration in the large intestine. Invasion through the gut wall into the peritoneal cavity, and bloodstream spread to other organs (primarily the liver) may occur.

Clinical manifestations
Intestinal amoebiasis, ranging from mild diarrhoea (often bloody) to severe colitis with systemic symptoms. Chronic infection can mimic inflammatory bowel disease (intermittent diarrhoea with abdominal pain and weight loss). Liver abscesses and, less commonly, lung and brain abscesses are found.

Laboratory diagnosis

Microscopy of freshly passed stools for the presence of trophozoites and cysts, or of trophozoites in abscesses; multiple stool specimens should be examined. It is important to distinguish cysts of *E. histolytica* from those of non-pathogenic amoebae. Serological tests (e.g. enzyme-linked immunosor-

bent assay [ELISA]) can be performed but are of limited value in endemic areas because antibodies persist for months or years. Serology is almost always positive in amoebic liver abscesses.

Treatment

Intestinal amoebiasis and hepatic amoebiasis are treated with metronidazole. Asymptomatic carriage may be treated with diloxanide.

Prevention

Stop food and water contamination with human faeces. Chlorination of water is not sufficient to kill cysts; filtration is required for cyst removal. Boiling water kills cysts.

Free-living amoebae

Infection may follow swimming in water contaminated with *Naegleria* and *Acanthamoeba* species.

Table 21.1 Classification of medically important parasites.

MEDICALLY IMPORTANT PARASITES	
Subkingdom	**Associated organisms**
Protozoa	Amoebae
	Ciliates
	Coccidia (sporozoa)
	Flagellates
Metazoa	Nematodes (roundworms)
	Cestodes (tapeworms)
	Trematodes (flukes)

Table 21.2 Medically important protozoa.

MEDICALLY IMPORTANT PROTOZOA	
Organism	**Principal infections**
Amoebae	
Entamoeba histolytica	Diarrhoea; dysentery; invades intestine, with ulceration; may spread to other organs, including liver and lungs
Naegleria and *Acanthamoeba*	Meningitis
Ciliates	
Balatidium coli	Diarrhoea
Coccidia (sporozoa)	
Cryptosporidium parvum	Diarrhoea
Isospora belli	Diarrhoea
Toxoplasma gondii	Glandular fever syndrome
	Congenital infections, with central nervous system defects; encephalitis in immunocompromised individuals
Plasmodium species	Malaria
Flagellates	
Giardia lamblia	Diarrhoea; malabsorption
Trichomonas vaginalis	Urogenital infections
Trypanosoma species	Sleeping sickness; Chagas' disease
Leishmania species	Visceral leishmaniasis
	Cutaneous leishmaniasis

Naegleria fowleri

This is a rare cause of meningoencephalitis. Infection follows swimming in fresh water contaminated with *N. fowleri*. The amoebae colonize the nasopharynx and then invade the central nervous system (CNS) via the cribriform plate. Cerebrospinal fluid (CSF) examination shows polymorphs and motile amoebae. Infection is frequently fatal, although some cases have been treated successfully with amphotericin.

Acanthamoeba species

This is a very rare cause of meningoencephalitis in immunocompromised patients. Eye infections, particularly keratitis and corneal ulceration, may result from contact lens-cleaning solutions contaminated with *Acanthamoeba* species (Plate 39). Diagnosis of eye infections may be made by Giemsa staining of corneal scrapings. Topical miconazole or propamidine isethionate, alone or combined with oral itraconazole, may be effective to prevent further visual deterioration. Corneal grafting may be needed to restore normal vision.

Ciliates

The only species pathogenic to humans is *Balantidium coli*, a common pathogen of pigs.

Pathogenesis
B. coli produces proteolytic cytotoxins that facilitate tissue invasion and intestinal mucosal ulceration.

Epidemiology
B. coli is found worldwide, with pigs and cattle as important reservoirs. Infection is via the faecal–oral route, with occasional outbreaks following contamination of water supplies. Person-to-person spread may occur. *B. coli* cysts are ingested and trophozoites formed, which invade the mucosa of the large intestine and terminal ileum.

Clinical manifestations
An asymptomatic carrier state can occur. *B. coli*

causes gastrointestinal infection; symptoms include abdominal pain and watery diarrhoea containing blood and pus. Mucosal ulceration is rarely followed by invasive infection.

Laboratory diagnosis
Microscopy for trophozoites and cysts in faeces.

Treatment
Tetracycline or metronidazole.

Coccidia

Classification
The Coccidia include *Cryptosporidium parvum*, *Isospora belli*, *Plasmodium* species, and *Toxoplasma gondii*. They undergo asexual (schizogony) and sexual (gametogony) reproduction and have a variety of hosts, including humans.

Cryptosporidium parvum (Chapter 34)

Structure and life cycle
Mature oocytes containing sporozoites are ingested. The sporozoites are released and attach to the intestinal epithelium and mature (schizogony). Sexual forms develop (gametogony) and a fertilized oocyst is produced which is passed in the faeces.

Laboratory diagnosis
Microscopy of faecal specimens by a modified acid-fast stain for the presence of oocysts.

Isospora belli

I. belli is a coccidian parasite that causes diarrhoea and malabsorption in immunocompromised patients. Treatment is co-trimoxazole for 1 month, with life-long suppressive therapy for AIDS patients.

Toxoplasma gondii

Classification
A coccidian parasite.

Structure and life cycle (Figure 21.1)

Cats are the definitive host. Asexual and sexual cycles result in oocyst formation. Oocysts develop into trophozoites which disseminate via the bloodstream, particularly to muscle and brain. Multiplication at these sites leads to tissue damage.

Epidemiology

Human infection with *T. gondii* is common, occurring worldwide. A wide variety of animals carry the organism. Humans become infected from:
• ingestion of undercooked meat contaminated with trophozoites
• ingestion of infected oocytes from cat faeces
• transplacental transmission (maternal infection or reactivation as a result of immunosuppression in pregnancy)
• cardiac transplantation; the recipient receives a heart containing toxoplasma cysts
• reactivation caused by immunosuppression, e.g. AIDS.

Clinical manifestations
• Most infections are asymptomatic.
• In immunocompetent hosts: glandular fever-like syndrome.
• In immunocompromised hosts: myocarditis, choroidoretinitis, meningoencephalitis.
• Congenital infection: choroidoretinitis, hydrocephalus and intracerebral calcification.

Diagnosis
• Serology (e.g. ELISA): by rising IgG antibodies in paired sera, or IgM antibodies to differentiate between active and previous infection. Serology in AIDS is not useful because IgG antibodies do not distinguish between reactivated and latent infection.

• PCR (polymerase chain reaction) detection of parasite DNA in blood, CSF or amniotic fluid may be useful.
• MRI (magnetic resonance imaging) of the brain shows characteristic multiple, dense, ring-enhancing lesions with contrast.

Histology

By examination of appropriate biopsies for cysts.

Treatment

Pyrimethamine, plus sulphonamide. Indefinite suppressive treatment in AIDS.

Prevention

Cooking meat. Avoiding contact with cat faeces.

Plasmodium

Plasmodia are coccidian or sporozoan parasites of erythrocytes with two hosts:
1 Mosquitoes, where sexual reproductive stages (gametogony) take place.
2 Humans or other animals, where asexual reproductive stages occur (schizogony).

There are four species of plasmodia (classified in Table 21.3), which vary in detailed life cycle and clinical features, and can be separated morphologically.

Life cycle
All plasmodia share a common life cycle (Figure 21.2) but with some important variations.
1 Anopheles mosquitoes bite humans and infectious plasmodia (sporozoites) are introduced into the bloodstream.
2 The sporozoites are carried to liver parenchymal

Table 21.3 Species of malaria infecting humans.

PLASMODIA INFECTING HUMANS		
Species	**Fever cycle (h)**	**Clinical condition**
Plasmodium falciparum	36–48	Malignant tertian malaria
Plasmodium malariae	72	Quartan malaria
Plasmodium ovale	36–48	Benign tertian malaria
Plasmodium vivax	36–48	Benign tertian malaria

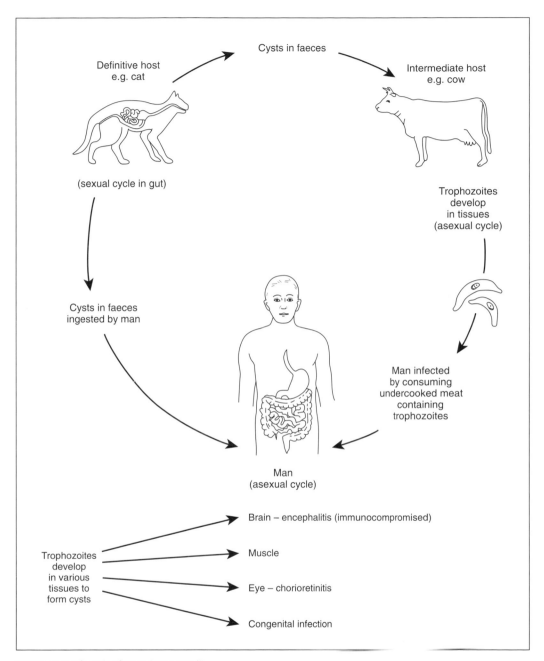

Figure 21.1 Life cycle of *Toxoplasma gondii*.

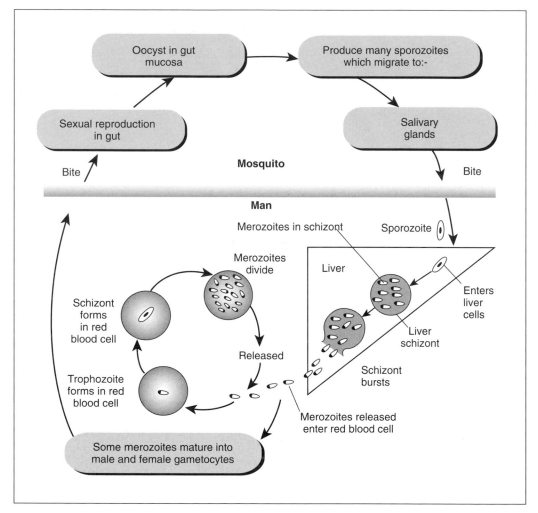

Figure 21.2 Life cycle of *Plasmodium falciparum.*

cells where asexual reproduction (schizogony) occurs in the liver to form merozoites (exoerythrocytic cycle, 7–28 days).

3 Liver hepatocytes rupture, liberating merozoites, which attach to and penetrate erythrocytes. Note: *P. vivax* and *P. ovale* have a dormant hepatic phase with sporozoites called hypnozoites that do not divide and can result in relapse years after the initial disease.

4 Asexual reproduction occurs in the erythrocytes; merozoites are released and infect further erythrocytes. Some merozoites develop within erythrocytes into male and female gametocytes. *P.*

falciparum may be associated with high levels of parasitaemia (up to 30% of circulating erythrocytes); lower levels (< 5%) are found with other species.

5 The mosquito bites a host and ingests mature male and female gametocytes.

6 The sexual reproductive cycle occurs in the mosquito's digestive system. Sporozoites form, migrate to salivary glands and are inoculated into a new host.

Epidemiology
• In tropical and subtropical areas plasmodia are

dependent on the correct conditions for breeding of anopheles mosquitoes. *P. falciparum* is responsible for > 80% of cases in tropical areas.

• In endemic areas, repeated infections or exposure results in relative immunity and less severe disease. Visitors to endemic areas are more severely affected.

• Transmission via contaminated blood transfusions or needle sharing can rarely occur.

Clinical features

• The incubation period is variable (10–40 days, but may be prolonged).

• Fever/sweats: symptoms are related to release of toxins when schizonts burst and so intervals between bouts of pyrexia are dependent on the erythrocyte cycle of the *Plasmodium* species (Table 21.3).

• Anaemia: this is caused by erythrocyte haemolysis. It is most severe with *P. falciparum* malaria and may result in haemoglobinuria ('blackwater fever').

• Cerebral malaria: high levels of parasitaemia associated with *P. falciparum* may result in erythrocyte debris blocking capillaries in the brain; the resultant hypoxia causes confusion and eventually coma, with a high mortality.

Laboratory diagnosis

Thick and thin blood films are taken. The typical morphology of the parasite within the erythrocytes allows the differentiation of the *Plasmodium* species.

Treatment

• Chloroquine is the treatment of choice for malaria caused by *P. vivax*, *P. ovale* and *P. malariae*. Supplementing treatment with primaquine is important to destroy the liver hypnozoite stages of *P. vivax* and *P. ovale*.

• Chloroquine is also the drug of choice for *P. falciparum* malaria, but chloroquine resistance is now very common in many areas of the world (e.g. South America, Central Africa, south-east Asia). Alternative drugs include quinine, mefloquine and arthemether with lumefantrine.

• Given the complexity of the epidemiology of

drug resistance, specialist advice should be taken when treating patients.

Prevention

This is important for travellers to endemic areas. Avoidance of mosquito bites is necessary (e.g. covering exposed limbs, and use of mosquito nets and repellants). Prophylactic anti-malarials may be taken. The exact regimen is dependent on whether resistance is present in the area being visited; examples include chloroquine, mefloquine and pyrimethamine with sulfadoxine (Fansidar).

Flagellates

Classification

Human pathogenic flagellates include *Giardia lamblia* (gastrointestinal tract), *Trichomonas vaginalis* (genital tract), *Trypanosoma* species (blood/tissues) and *Leishmania* species (blood/tissues).

Giardia lamblia (see Chapter 34)

Pathogenesis and life cycle

Cysts are ingested and gastric acid stimulates the release of trophozoites in the small intestine, which then multiply by binary fission. Trophozoites attach to the intestinal villi by a sucking disc (Plate 40). Inflammation of the epithelium may occur, but systemic invasion is rare. Trophozoites divide by binary fission. Cyst formation occurs as the organisms move through colon.

Epidemiology

Giardia occurs worldwide. Transmission is via ingestion of contaminated water or food, or direct person-to-person spread via the faecal–oral route.

Clinical manifestations

Asymptomatic carriage is common. Active infection causes diarrhoea and malabsorption occasionally. Chronic infection may cause failure to thrive in children.

Laboratory diagnosis

Stool microscopy for cysts and trophozoites (see Plate 40). If clinical suspicion is high and stool

Table 21.4 Clinical comparison of the types of malaria.

TYPES OF MALARIA				
Characteristic	*P. vivax*	*P. ovale*	*P. malariae*	*P. falciparum*
Incubation period (days)	10–17 Sometimes prolonged for months to years	10–17	18–40	8–11
Duration of untreated infection	5–7 years	12 months	20+ years	6–17 months
Anaemia	++	+	++	++++
Central nervous system involvement	+	±	+	++++
Renal involvement	±	+	++++	+

± to ++++, less likely to very common.

examination is negative, duodenal aspiration or biopsy may be helpful.

Treatment
Metronidazole.

Prevention
The cysts resist routine levels of chlorination. Water can be decontaminated by boiling or filtration.

Trichomonas vaginalis

Structure
T. vaginalis is a pear-shaped protozoa with four flagella. No cyst stage has been recognized.

Epidemiology
It has a worldwide distribution, with sexual intercourse being the primary method of spread.

Pathogenesis/clinical conditions
T. vaginalis is a parasite of the human urogenital system; there are no animal reservoirs. Infection results in a watery vaginal discharge, although extensive inflammation and erosion of the epithelium can occur. Males are usually asymptomatic carriers; however, occasionally, urethritis and prostatitis can occur.

Laboratory diagnosis
This is by microscopy of vaginal or urethral discharge for trophozoites or isolation in special media.

Treatment
Metronidazole for affected patient and sexual partner(s).

Trypanosoma species

Trypanosomes are haemoflagellates living in the blood and tissue of human hosts. The life cycle involves two hosts: blood-sucking insects and mammals. It causes two diseases:
1 African trypanosomiasis (sleeping sickness) caused by *T. brucei gambiense* and *T. brucei rhodesiense* and transmitted by the tsetse fly.
2 American trypanosomiasis (Chagas' disease) caused by *T. cruzi*, and transmitted by the reduviid bug.

Trypanosoma brucei gambiense (West African sleeping sickness)
Life cycle
1 The infective stage (trypomastigote) is present in the salivary glands of the tsetse fly. The trypomastigote has a flagellum and an undulating membrane along the length of the body allowing motility.
2 Trypomastigotes enter the host via insect bites and reach the lymphatic system, blood and CNS. Non-flagellate forms (amastigotes) develop in some tissues (e.g. heart and muscle).

3 Tsetse flies feed on the infected host and take in trypomastigotes which multiply in the intestine; epimastigotes subsequently form in the salivary glands and develop into infective trypomastigotes.

4 The tsetse fly remains infective for life.

Epidemiology

Limited to west and central Africa. Domestic animals and asymptomatic humans are the main reservoirs.

Clinical condition

The incubation period is a few days to several weeks. A nodule or chancre can form at the site of the bite, which usually resolves spontaneously. The parasite enters the bloodstream and results in lymphadenopathy, irregular fever and myalgia. CNS involvement (sleeping sickness) may occur with lethargy and encephalitis, leading to convulsions, hemiplegia, coma and occasionally death.

Diagnosis

Microscopy of thick and thin blood films, lymph node aspirates and CSF for trypomastigotes; serology (e.g. ELISA).

Treatment

Suramin; melarsoprol if there is CNS involvement.

Prevention

Vector eradication. insect repellents, and thick wrist- and ankle-length clothing.

Trypanosoma brucei rhodesiense (East African sleeping sickness)

Life cycle

The life cycle is similar to that of *T. brucei gambiense*, with trypomastigote and epimastigote stages. Transmission is by the tsetse fly.

Epidemiology

It is found in east Africa. Domestic and game animals are the main reservoirs. Asymptomatic human carriers are not normally a source of infection.

Clinical conditions

It has a shorter incubation than *T. brucei gambiense*. The illness is more severe and progresses rapidly and may involve the heart, kidney and CNS. Mortality is greater than for the west African form.

Diagnosis, treatment and prevention

As above.

Trypanosoma cruzi (Chagas' disease)

Life cycle

- The life cycle is similar to other trypanosomes, except, when the infected bug bites humans, trypomastigotes are released simultaneously in the faeces; these enter the wound after scratching.
- Spread is via the lymphatic and blood systems and results in invasion of many organs, including the liver, heart, muscles and brain.
- Within host cells, trypomastigotes transform into amastigotes, which multiply by binary fission and form either further amastigotes or trypomastigotes; the latter are ingested by a feeding insect, multiply in the intestine and are then passed in the faeces.

Epidemiology

T. cruzi is found in North, Central and South America. Reservoirs are humans and domestic animals.

Clinical disease

Chagas' disease may be asymptomatic, acute or chronic. A painful nodule may form at the bite. Acute infection results in high fever, erythematous rash, oedema and myocarditis. Chronic disease may develop years after the initial infection. Organisms proliferate in various organs, including the brain, spleen, liver, heart and lymph nodes, resulting in lymphadenopathy, hepatosplenomegaly, myocarditis and cardiomegaly. Granulomata and cysts may form in the brain. Chronic gastrointestinal disease produces symptoms resembling achalasia or Hirschsprung's disease. Cardiac disease is the most common presentation, with congestive cardiac failure. Transplacental transmission may occur in chronically infected women. Transmission by blood transfusion has occurred.

Laboratory diagnosis

Microscopy of biopsies of affected tissues; thick blood films for trypomastigotes in the acute phase; serological tests (e.g. ELISA) may produce false positives with leishmaniasis; PCR amplification of parasite DNA from blood or body fluids.

Treatment

Nifurtimox; supportive measures for complications.

Prevention

Insecticides; avoidance of insect bites, sleeping under netting; improvement in living conditions. Blood for transfusion should be screened for antibodies to *T. cruzi*. Adding gentian violet to blood for transfusion will kill parasites.

Leishmania species

Leishmania species are obligate, intracellular parasites transmitted to mammalian hosts by sandflies. An estimated 12 million people are infected worldwide.

Classification

There are three main species that produce human disease: *Leishmania donovani*, *L. tropica* and *L. braziliensis*. These species are associated with various clinical syndromes (Table 21.5).

Structure

Leishmaniae are flagellated protozoa.

Life cycle

1 A promastigote stage is present in the saliva of infected sandflies and is inoculated into the bite site.
2 Promastigotes lose their flagella, enter the amastigote stage, are engulfed by tissue macrophages, and carried via the reticuloendothelial system to the bone marrow, spleen and liver. Amastigotes multiply, resulting in tissue damage.
3 Amastigotes are taken up by the sandfly during feeding, develop into promastigotes, multiply in the mid-gut and then migrate to the salivary glands.

Visceral leishmaniasis

This is caused by *L. donovani*. The infection occurs in India, China, southern Russia, Africa, the Mediterranean basin and Latin America.

Clinical conditions

It can be asymptomatic. Fever, weight loss and diarrhoea are present, with hepatosplenomegaly and renal involvement. Darkening of the skin, mainly the face and hands (kala-azar), malabsorption and anaemia may occur. Most infections resolve without therapy. Those who recover from infection, are unlikely to have reinfection, unless their cell-mediated immunity is suppressed (e.g. AIDS).

Diagnosis

Histological examination of appropriate tissue biopsies; bone marrow or lymph nodes for amastigotes; serology. PCR amplification or

Table 21.5 Syndromes caused by *Leishmania* species.

LEISHMANIA SPECIES AND INFECTIONS				
Organism	Reservoir	Clinical condition	Location	Average incubation period
L. donovani	Humans or dog	Visceral leishmaniasis (kala-azar)	Africa Asia	3 months
L. tropica	Humans or dog	Cutaneous leishmaniasis (oriental sore)	Africa Asia	1–2 months
L. major	Rodents		Mediterranean	
L. braziliensis	Sloths and related species	Mucocutaneous leishmaniasis	Central and South America	A few weeks to months

nucleic acid-based detection of parasite DNA increases sensitivity

Treatment
Sodium stibogluconate or amphotericin B .

Cutaneous leishmaniasis

Caused by *L. tropica* and *L. major* complexes in southern Europe, Asia and Africa, and by *L. mexicana* complex in Latin America

Clinical conditions
Cutaneous disease with a papule at the bite site, followed by necrosis of epidermis and ulceration. Infections remain localized, although lesions may be multiple.

Diagnosis
Histological examination for amastigotes or DNA detection in smears or biopsy material; serology.

Treatment
Sodium stibogluconate.

Mucocutaneous leishmaniasis

Caused by *L. braziliensis* complex in Latin America.

Clinical conditions
Infection mainly involves the mucous membranes of the upper respiratory tract (palate, nose) and related tissue, with ulceration, tissue destruction and a resulting disfigurement.

Diagnosis
Histological examination or DNA detection in biopsies; serology.

Treatment
Sodium stibogluconate or amphotericin B.

Chapter 22

Parasitology: metazoa (helminths)

Nematodes

The most common nematodes of medical importance include *Ascaris lumbricoides*, *Enterobius vermicularis*, *Toxocara canis* and *Toxocara cati* (Table 22.1).

Ascaris lumbricoides (roundworm)

Structure and life cycle (Figure 22.1)
- Adult worms are cylindrical; the female is up to 35 cm long and the male up to 30 cm long.
- After ingestion of an infective egg, larval worms are released into the duodenum, which penetrate the intestinal wall to reach the bloodstream, passing via the liver and heart to the pulmonary circulation. These larval worms pass into the alveoli, migrate via the bronchi, trachea and pharynx, are swallowed and return to the small intestine.
- Male and female worms mature and mate in the intestine; up to 200 000 eggs per day are produced by the female and passed in stools.

Epidemiology
Ascaris lumbricoides is found throughout the world, but particularly in areas of poor sanitation where faecal–oral transmission may occur. Eggs can survive for long periods (several years), making *Ascaris* the most common pathogenic helminth worldwide.

Clinical manifestations/complications
Victims are asymptomatic when the worm load is light. Vomiting and abdominal discomfort may occur. Complications include:
- pneumonitis (following larval migration in the lungs)
- intestinal obstruction with mature worms (children)
- malnutrition in children
- perforation of the intestine and hepatic abscesses (rare).

Laboratory diagnosis
Stool examination for the presence of fertilized or unfertilized eggs. An adult worm is occasionally passed in faeces or vomit. Larvae are sometimes found in sputum in the lung migration phase.

Treatment and prevention
Treatment is with mebendazole, albendazole or piperazine. Surgical or endoscopic relief of intestinal obstruction. Control is by improved sanitation and avoidance of food contaminated with human faeces.

Enterobius vermicularis (pin worm)

Structure and life cycle (Figure 22.2)
- The female worm (10 mm × 0.4 mm) has a pointed tail; the male worm (3 mm × 0.2 mm) has a

Table 22.1 Medically important nematodes.

MEDICALLY IMPORTANT NEMATODES	
Organism	**Notes on disease/infection**
Ascaris lumbricoides (roundworm)	Larval worms migrate via bronchi; adult worms may cause intestinal obstruction in children
Enterobius vermicularis (pin- or threadworm)	Perianal itching
Trichinella spiralis	Periorbital oedema; myalgia; fever
Toxocara canis	Larvae migrate to liver, lungs and eyes (visceral larva migrans)
Trichuris trichiura	Anaemia, intestinal irritation

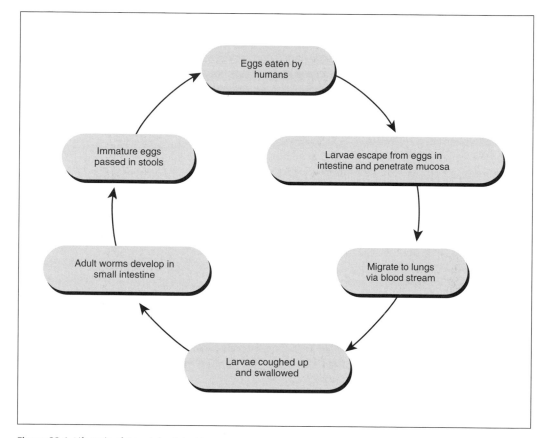

Figure 22.1 Life cycle of *Ascaris lumbricoides*.

curved end. Infection follows ingestion of an embryonated egg.

• Larvae hatch in the small intestine and migrate to the large intestine where they mature into adults in approximately 1 month.

• The male fertilizes the female which migrates to the perianal area and lays eggs, which become infective within hours.

• Perianal irritation leads to scratching; eggs contaminate hands, particularly nails, resulting in autoinfection or transmission to a new host.

Epidemiology

This is a common parasitic infection, with 500 million cases worldwide, mainly in temperate

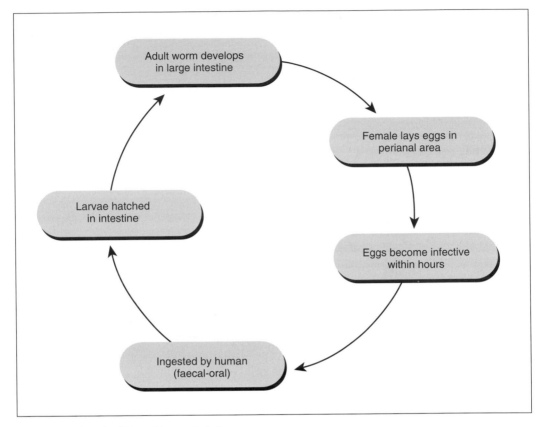

Figure 22.2 Life cycle of *Enterobius vermicularis.*

regions. Person-to-person or environment-to-person spread is responsible for transmission.

Clinical manifestations and complications
Individuals are frequently asymptomatic. Patients allergic to worm secretions may experience perianal pruritus. In heavily infected females, worm migration into the vagina may occur.

Laboratory diagnosis
By microscopy; the best method is a Sellotape slide applied around the perianal region to pick up eggs. Adult worms may be found in stool specimens.

Treatment and prevention
Mebendazole is given to treat the entire family simultaneously. Reinfestation may occur, because eggs may survive in the environment for 3 weeks. Improved personal hygiene is advised, including cutting of fingernails and regular washing of bedclothes and towels.

Toxocara canis and *Toxocara cati*

Structure and life cycle
• *T. canis* and *T. cati* are parasites of dogs and cats, respectively. Male and female worms develop in the intestine and produce eggs that are ingested by other dogs or cats and the cycle continues.
• Humans are an unintentional host; on ingestion, the eggs hatch into larval forms, which can penetrate the intestinal mucosa and migrate to various organs, particularly the lung, liver and eyes. Larvae are unable to develop further and granuloma formation results.

Epidemiology
It is associated with infected dogs and cats; worldwide distribution.

Clinical manifestations and complications
Toxocariasis may result in pneumonitis, hepatosplenomegaly, eosinophilia and retinitis.

Laboratory diagnosis
Serology by ELISA. Biopsies of the liver or other affected organs may show eosinophilic granulomata.

Treatment
Often unnecessary; albendazole or thiabenazole; steroids are also used in severe cases. Laser photocoagulation may kill larvae in the eye.

Prevention
Young pets should be treated routinely ('deworming'); avoid contact with animal faeces.

Trichuris trichiura (whipworm)

Structure and life cycle
- Ingested eggs develop into the larvae in the small intestine which then migrate to the large intestine and mature into adult worms. The fertilized female worm produces eggs.
- Eggs are excreted and mature in the soil.

Epidemiology
Trichuris trichiura has a wide distribution in the tropics and subtropics, particularly in areas with poor sanitation. No animal reservoir has been identified.

Clinical manifestations and complications
Patients are usually asymptomatic, but worm infection may produce abdominal distension and diarrhoea with weight loss; anaemia, eosinophilia, bloody diarrhoea and anal prolapse are seen occasionally.

Laboratory diagnosis
Stools are examined for characteristic eggs.

Treatment
None for asymptomatic or light infections. Mebendazole is the drug of choice for more severe infections.

Prevention
Good personal hygiene and adequate sanitation can help to prevent infection.

Ancylostoma duodenale and Necator americanus (hookworm)

Structure and life cycle
- Infective larvae penetrate intact skin, enter the circulation and are carried to the lungs; after maturation, larvae are expectorated and swallowed, and develop into adult worms in the small intestine.
- The adult worms lay eggs which are excreted and, in hot humid conditions, develop rapidly into infective larvae that may infect a new host by penetration of exposed skin, particularly bare feet.

Epidemiology
Found primarily in subtropical and tropical regions.

Clinical manifestations and complications
- There is a rash at the entry site; occasionally pneumonitis occurs during the larval migration.
- Adult worms produce gastrointestinally related symptoms, including vomiting and bloody diarrhoea.
- Chronic anaemia may result from blood loss, particularly in malnourished individuals.

Laboratory diagnosis
Stool microscopy for eggs.

Treatment and prevention
Mebendazole for treatment; preventive measures include improved sanitation and wearing of shoes to prevent soil contact in endemic areas.

Strongyloides stercoralis

Structure and life cycle
- *Strongyloides stercoralis* is similar to hookworms,

except that the eggs hatch into larvae (Plate 41) in the intestinal mucosa before being passed in the faeces. The larvae mature directly into infective larvae that are able to penetrate skin, or enter a non-parasitic cycle outside the human host.

• Larvae may also reinfect the host by penetrating the intestinal mucosa (autoinfection); this is more common in immunocompromised patients.

• Infection may be quiescent for up to 30 years and then reactivate when the immune system is compromised (e.g. organ transplantation), resulting in disseminated infection.

Epidemiology
They are found in warm, moist environments, similar to hookworm.

Clinical manifestations
Pneumonitis, diarrhoea, malabsorption and gut ulceration may occur. Immunosuppressed patients may develop severe infection affecting multiple organs.

Laboratory diagnosis
Stool microscopy for larvae; sampling of proximal small intestine with string capsules or by endoscopic aspiration may be needed in low-level infections. In disseminated infection, filariform larvae may be found in stool, duodenal contents, sputum and bronchial washings. Serology; 85% sensitive in uncomplicated infection, but there are false positive results with other intestinal nematode infections.

Treatment
Thiabenazole—repeated courses may be necessary; ivermectin may be effective.

Prevention
Adequate sanitation.

Trichinella spiralis

Structure and life cycle
• Human infection occurs accidentally because this parasite is found primarily in carnivorous animals. Humans may become affected after ingestion of encysted larvae in undercooked meat, particularly pork.

• Larvae are released in the small intestine where they mature into adult worms that produce further larvae. The larvae penetrate the intestinal wall and are released into the circulation, becoming encysted in various tissues, particularly striated muscle.

• The life cycle is completed in animals who are infected after ingestion of striated muscle containing encysted larvae.

Epidemiology
Worldwide distribution; associated with eating pork or the meat of wild carnivores.

Clinical manifestations
Trichinellosis presents with muscle pain, tenderness and periorbital oedema. Encephalitis and pneumonitis may occur. Infection is occasionally fatal, usually as a result of myocarditis.

Laboratory diagnosis
Eosinophilia, the presence of encysted larvae in implicated meat or biopsied muscle from the patient, and serology can be used.

Treatment
Mebendazole kills adult worms but has no effect on encysted larvae; steroids are used in severe allergic reactions or myocarditis.

Prevention
Meat should be cooked thoroughly.

Wuchereria bancrofti and Brugia malayi (filarial worms)

These are transmitted by mosquitoes.

Structure and life cycle
• Immature larvae are introduced after a bite from an infected mosquito; they migrate via the lymphatics to regional lymph nodes.

• Adult worms mature in the lymphatics, fertilization occurs and larval microfilariae are produced. These enter the circulation and are ingested by feeding mosquitoes.

- In the mosquito, larvae migrate via the stomach to the mouthpiece to complete the cycle.

Epidemiology
The worms are found in tropical and subtropical Africa, Asia and South America.

Clinical manifestations and complications
Early signs include flu-like illness, lymphangitis and lymphadenopathy. Chronic infection results in blockage of the lymphatics with peripheral oedema, and eventually limb fibrosis (elephantiasis).

Laboratory diagnosis
By detection of microfilariae in blood films (taken at night as microfilariae show nocturnal periodicity) and serology.

Treatment
Diethylcarbamazine.

Prevention
Mosquito control; avoiding bites.

Loa loa

Structure and life cycle
- *Loa loa* is similar to *Wuchereria bancrofti*. The insect vector is *Chrysops*, the mango fly.
- In humans, larvae mature and migrate subcutaneously. Fertilization results in microfilariae, which pass into the bloodstream and are ingested by feeding flies.

Epidemiology
Loa loa infection is found in tropical Africa.

Clinical manifestations
There is swelling where the worms migrate subcutaneously. Adult worms may be seen migrating under the conjunctiva, hence the name 'eye-worm'.

Laboratory diagnosis
By detection of microfilariae in blood and serology.

Treatment and prevention
Treatment is with diethylcarbamazine or ivermectin, and by surgical removal of localized worms. Prevention is by protection from insect bites.

Onchocerca volvulus

Infection follows a bite from black flies which breed near fast-flowing rivers. Larvae develop into adult worms in subcutaneous nodules. *Onchocerca volvulus* is found in Africa and central southern America. Chronic infection results in atrophic skin and may cause blindness as a result of microfilariae migrating to the eyes. Diagnosis is by skin biopsy or slit-lamp examination of the anterior chamber of the eye for microfilariae. Killing black fly larvae is an important control measure. Bites can be reduced by clothing and insect repellents. Annually administered ivermectin effectively controls disease and may decrease transmission of the parasite. Ivermectin is also used for treatment.

Cestodes

Tapeworms are ribbon like and up to 30 feet in length. A head or scolex has suckers with a crown of hooklets, facilitating attachment. Segments of tapeworms are called proglottids. All tapeworms are hermaphrodites, with both male and female reproductive organs present in each segment. Food is absorbed from the host intestine through the body wall. Most tapeworms have complex life cycles, with intermediate hosts. Clinically important cestodes are listed in Table 22.2.

Taenia solium

Structure and life cycle (Figure 22.3)
Humans ingest pork containing larvae (cysticerci) and the larvae develop in the small intestine into the adult form. The worm produces proglottids which mature sexually, producing eggs that pass in the faeces. Pigs ingest the eggs and larval forms develop that disseminate via the circulation to muscle to produce cysticerci.

Table 22.2 Clinically important cestodes.

CLINICALLY IMPORTANT CESTODES			
Cestode	Common name	Reservoir for larvae	Reservoir for adult worm
Taenia solium	Pork tapeworm or cysticercosis	Pigs	Humans
Taenia saginata	Beef tapeworm	Cattle	Humans
Diphyllobothrium latum	Fish tapeworm	Crustacea, fish	Humans, cats, dogs
Echinococcus granulosus	Hydatid cyst	Humans	Canines
Hymenolepis nana	Dwarf tapeworm	Rodents, humans	Rodents, humans

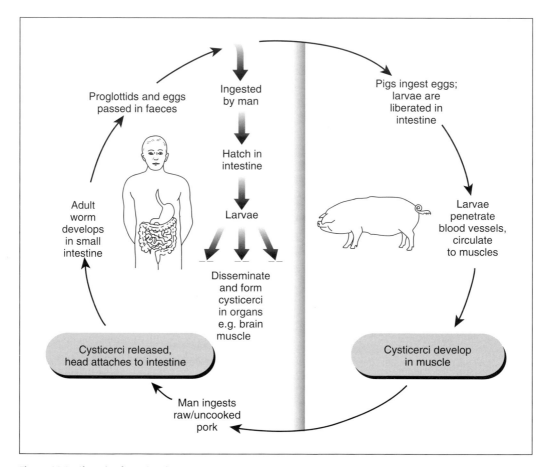

Figure 22.3 Life cycle of *Taenia solium*.

Epidemiology
Infection is related to eating undercooked pork. It is common in Africa, India, south-east Asia and Central America.

Clinical manifestations
Abdominal discomfort and diarrhoea.

Diagnosis
Stool microscopy for proglottids and eggs; serology.

Treatment and prevention
Treatment is with praziquantel or albendazole. Pork should be cooked thoroughly and improved sanitation is important for prevention.

Cysticercosis

Humans may become infected with the larval stage of *T. solium* after ingestion of eggs from human faeces. Larvae penetrate the intestinal wall, enter the circulation and are carried to muscles, the brain, lungs and occasionally eyes. The inflammatory response to the presence of these larvae in human tissues eventually results in calcification. Infection is often subclinical but may present with neurological or ophthalmic manifestations.

Diagnosis

Muscle or brain radiograph may show calcified cysts; computed tomography (CT) or MRI is often diagnostic; histology; serology (ELISA, immunoblot).

Treatment

Praziquantel, but may not be necessary if cysts are not viable. Steroids and anticonvulsants may be necessary to control symptoms if CNS cysts are degenerating. Obstructive hydrocephalus may need surgery.

Taenia saginata

Structure and life cycle

The life cycle is similar to *Taenia solium*, with infection resulting from ingestion of cysticerci present in insufficiently cooked beef. Human faeces contaminated with eggs are ingested by cattle and develop in beef.

Epidemiology

Worldwide distribution.

Clinical manifestations

Abdominal discomfort.

Laboratory diagnosis

Proglottids and eggs are detected by stool microscopy.

Treatment and prevention

Treatment with praziquantel or albendazole. Improved sanitation aids prevention. Beef should be cooked thoroughly.

Diphyllobothrium latum (fish tapeworm)

Structure and life cycle

There are two intermediate hosts: freshwater crustaceae and freshwater fish. Infection occurs in humans after ingestion of infected freshwater fish. The adult worm develops in the intestine, attaching to the mucosa via lateral grooves. Proglottids develop and produce eggs. In fresh water the eggs develop into larvae, which infect crustaceae. Fish eat the crustaceae and larvae develop that are infective for humans.

Epidemiology

Infection occurs in temperate climates, particularly Scandinavia.

Clinical manifestations

Usually asymptomatic. Mild diarrhoea and occasionally intestinal obstruction. Megaloblastic anaemia and vitamin B_{12} deficiency occur.

Treatment and prevention

Praziquantel is the drug of choice. Avoid eating raw and undercooked fish.

Echinococcus granulosus (dog tapeworm)

Structure and life cycle

• The adult tapeworms are found in canines, normally dogs or foxes. In the canine intestine the adult tapeworms produce eggs, which are passed in the faeces.

• Herbivores, the normal intermediate hosts, ingest the eggs and larvae develop in various organs, resulting in cyst formation. Carnivores become infected after consuming contaminated carcasses.

• Human infection occurs when humans become the accidental intermediate host. Eggs are ingested, and larvae develop and penetrate the intestinal wall; hydatid cysts develop, principally in the liver. These cysts accumulate fluid and enlarge over several years; a germinal layer surrounds the cyst, which produces immature tapeworm heads.

111

Epidemiology

Echinococcus granulosus is associated with sheep farming in Europe, Australasia, South America and the USA. Human infection follows ingestion of water or vegetation contaminated with eggs.

Clinical manifestations

Cysts in the liver grow over many years, resulting in hepatomegaly and jaundice. Cysts in the lung, bone and brain may result in various clinical presentations.

Diagnosis

CT or ultrasonography may be diagnostic; microscopy of cyst contents; serology is highly specific.

Treatment and prevention

Treatment is by resection of cysts or aspiration under CT guidance with the scolicidal agent albendazole. Avoid ingestion of contaminated food.

Hymenolepsis nana

This is a common, often asymptomatic infection occurring particularly in Asia. *Hymenolepsis nana* is a small (2–5 cm) worm. Eggs are produced in the small intestine and may reinfect the same host or spread to other hosts by the faecal–oral route. Heavy infection may result in diarrhoea, abdominal discomfort and weight loss. Treatment is with praziquantel.

Trematodes

Except for schistosomes, trematodes (flukes) are flatworms with oral and ventral suckers. Most flukes are hermaphrodites, containing both male and female reproductive organs in a single body. Schistosomes differ in having separate male and female worms.

All flukes require a reservoir host and intermediate host for completion of their life cycle. The intermediate hosts are molluscs, in which the sexual reproduction cycle occurs. Some flukes need a further intermediate host. The medically important trematodes are summarized in Table 22.3.

Fasciola hepatica (sheep liver fluke)

Structure and life cycle

- *F. hepatica* is a parasite of herbivores (sheep, cattle) and occasionally humans.
- Human infection starts with ingestion of watercress or similar vegetation contaminated with encysted metacercariae, which excyst in the duodenum. The larval flukes migrate through the duodenal wall, across the peritoneal cavity and into the bile ducts via the liver. They mature into adult worms and produce eggs that are passed in faeces.
- Snails become infected and act as intermediate hosts. Cercariae (infective larvae) develop and contaminate watercress where they encyst to form metacercariae.

Table 22.3 Medically important trematodes (flukes).

MEDICALLY IMPORTANT TREMATODES				
	Common name	**Intermediate host**	**Vector**	**Reservoir host**
Fasciola hepatica	Sheep liver fluke	Snail	Aquatic plants (e.g. watercress)	Sheep, cattle
Opisthorchis sinensis humans (formerly *Clonorchis sinensis*)	Chinese liver fluke	Snail, freshwater fish	Uncooked fish	Dogs, cats,
Paragonimus westermani	Lung fluke	Snail, crayfish	Uncooked crabs, crayfish	Humans
Schistosoma species	Blood flukes	Snail		Humans

Epidemiology
F. hepatica occurs worldwide, especially in sheep-rearing areas.

Clinical manifestations
Migration of the larval worm via the liver causes hepatomegaly. Worms in the bile duct may cause biliary obstruction and eventually cirrhosis.

Laboratory diagnosis
Stool microscopy for eggs.

Treatment and prevention
Praziquantel. Avoid eating watercress grown in areas frequented by sheep and cattle.

Opisthorchis sinensis (Chinese liver fluke; formerly known as Clonorchis sinensis)

Structure and life cycle
O. sinensis has a similar life cycle to F. hepatica except there are two intermediate hosts. Eggs are eaten by a snail and develop into adult worms; free-swimming cercariae leave the snail and enter freshwater fish by penetrating under scales where they develop into metacercariae. Humans become infected by eating contaminated raw fish. Larval forms penetrate the duodenum, migrate to the bile duct and develop into adult worms.

Epidemiology
O. sinensis is found in China, Japan, Korea and south-east Asia.

Clinical manifestations
People are usually asymptomatic, but severe infections can result in hepatomegaly, biliary obstruction and jaundice.

Laboratory diagnosis
Stool microscopy for eggs.

Treatment and prevention
Praziquantel is the drug of choice. Biliary obstruction may need surgery. Avoid eating uncooked fish. Improved sanitation is important in prevention.

Paragonimus westermani (lung fluke)

Structure and life cycle
• Embryonated eggs in water produce miracidia, which penetrate snails and develop into cercariae. The latter infect crabs and crayfish.
• Humans become infected by eating infected crabs or crayfish. The larvae hatch in the duodenum, migrate through the intestinal wall, cross the peritoneal cavity and eventually reach the lungs via the diaphragm.
• The worms reside within fibrous cysts in the lungs and produce eggs that then appear in the sputum or, when swallowed, in faeces.

Epidemiology
P. westermani is found in Asia, Africa, India and South America.

Clinical manifestations
Adult flukes in lungs may result in cough and chest pain; lung fibrosis may develop. Larval migration can also cause disease in other organs, particularly the brain.

Laboratory diagnosis
Microscopy of sputum and faeces for eggs and serology.

Treatment and prevention
Praziquantel. Avoid eating uncooked freshwater crabs and crayfish; improve sanitation.

Schistosomes

Schistosomes associated with human infection (schistosomiasis) include *Schistosoma mansoni*, *S. japonicum* and *S. haematobium* (Table 22.4). Schistosomiasis is an important infection with over 200 million people infected worldwide.

Structure and life cycle (Figure 22.4)
• Schistosomes have separate male and female worms.
• Infective cercariae penetrate the skin, enter the circulation, and develop into male and female worms in the veins of the intestine (*S. mansoni*,

Table 22.4 Diagnosis of schistosome infection.

SCHISTOSOMIASIS: EPIDEMIOLOGY AND LABORATORY DIAGNOSIS			
Schistosoma species	Location in human	Epidemiology	Laboratory diagnosis
S. mansoni	Inferior mesenteric vein	Endemic in Africa, Middle East and South America	Stool for eggs with lateral spine Serology
S. japonicum,	Superior and inferior mesenteric vein	China and Japan	Stool for eggs
S. mekongi		Laos and Cambodia	Rectal biopsy for eggs Serology
S. haematobium	Inferior pelvic system	Egypt and Middle East	Urine for eggs with terminal spine Bladder biopsy for eggs Serology

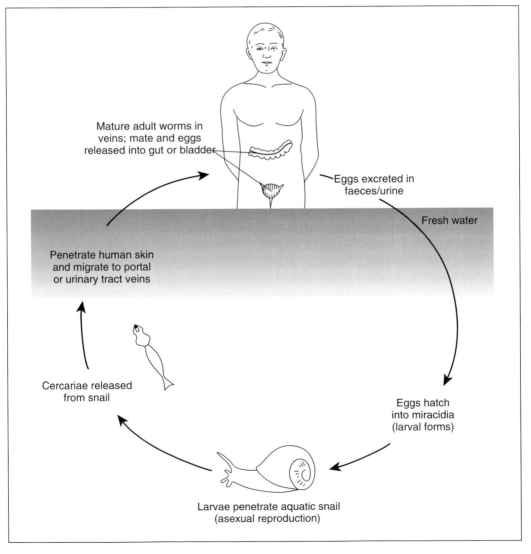

Figure 22.4 Life cycle of schistosomes.

S. japonicum) or bladder (*S. haematobium*). The female worm produces eggs that penetrate the intestinal or bladder wall and are passed in faeces or urine. Ova reach fresh water, hatch into larvae and infect snails (Plate 42). An asexual cycle results, with formation of cercariae that are released into water.

Clinical manifestations
• Skin penetration may result in local dermatitis.
• Egg release is associated with haemorrhage, normally minor, into the bladder (haematuria) or intestine (blood in faeces).
• Egg entrapment in the bladder or intestinal wall may result in a chronic inflammatory reaction with subsequent fibrosis:
(1) fibrosis and calcification of the bladder wall; obstruction of ureters, resulting in hydrone-phrosis and renal failure; malignant change in bladder.
(2) fibrosis of the portal vein, resulting in portal hypertension.
• Egg deposition may occur in other organs, e.g. the liver (fibrosis, portal hypertension), lungs and brain.

Laboratory diagnosis
Table 22.4 gives details of laboratory diagnosis of schistosomes.

Treatment and prevention
Praziquantel for all species. Oxamniquine is effective only against *S. mansoni*. Prevention is by improvement of sewage disposal; education (e.g. cover feet in endemic areas), and the use of molluscicides.

Chapter 23

Host–parasite relationships

Introduction

The development of infection depends on the interaction of microorganisms ('the parasite') and humans ('the host'). Microorganisms vary in their ability to produce infection (pathogenicity); in many microorganisms, this ability relates to their *virulence factors*. The virulence factors may also affect transmission of microorganisms. The host immune response to infecting organisms determines whether or not an infection is asymptomatic, and its outcome. In some instances, this response is responsible for the major clinical manifestations of infection, e.g. the cell-mediated immune response ('hypersensitivity') to tuberculosis (TB).

Ecology

Microorganisms have developed characteristics that allow them to survive and grow in a wide variety of environments; they are found in virtually every ecological niche. They play a central role in global ecology, both providing an important food source and facilitating the degradation of organic matter.

- Microorganisms that live on inanimate organic material are described as *saprophytes*.
- Microorganisms that grow in or on a living host, obtaining nutrients from the host, are referred to as *parasites*; they may be beneficial to the host (*symbiosis*), e.g. in the stomach of ruminants they aid in the degradation of cellulose—a vital step in the nutrition of such animals.

Normal flora

Parasitic microorganisms associated with mammals are often referred to as normal flora or commensals. They are important in preventing infection; a reduction in the normal flora (e.g. by antibiotic therapy) may result in overgrowth of potential pathogens.

Commensals are well adapted to their environment, e.g. the dry, acidic conditions found on the skin, and can multiply without causing host damage. However, commensal bacteria can cause infection, normally when some aspect of host defence is compromised, e.g.:

- *Staphylococcus epidermidis*, a skin commensal, is a cause of intravascular catheter infections (see p. 220, Plate 63).
- Viridans streptococci (oral commensal bacteria) may cause bacterial endocarditis (see p. 228).

Pathogenic microorganisms

Microorganisms that frequently cause infections are referred to as *pathogenic*. However, they may be present on the host without resulting in infection, e.g.:

- *Neisseria meningitidis*, an important cause of

meningitis, is a commensal of the pharynx of about 5% of adults.

• *Streptococcus pneumoniae*, the most frequent cause of bacterial pneumonia, is a pharyngeal commensal in many individuals.

• *S. aureus*, an important cause of skin infection and deep abscesses, is a nasal commensal of about one-third of individuals.

Infection

• Infection describes the clinical manifestations that occur when a microorganism invades a host.

• When the manifestations are minor or imperceptible, infections are often termed 'subclinical' or 'asymptomatic'.

• *Latent infection* results when pathogens persist in the body without evoking a clinical response; periodically, overt infections occur after a change in the patient's immune state (e.g. cytomegalovirus infection in transplant recipients; herpetic cold sores).

• *Chronic* or *persistent infections* occur when pathogens are not eradicated completely and continue to evoke a clinical response (e.g. leprosy, helminth infections, HIV).

Microbial strategies

Microorganisms have developed a variety of mechanisms (virulence factors) to combat host defences and promote transmission to new hosts.

Transmission

The ability to survive outside the host is an important factor in transmission of organisms. Upper respiratory tract viruses (e.g. rhinoviruses) survive poorly outside the host and successful transmission relies on the production of large quantities of infectious particles. Other organisms can survive outside the host, e.g. *Mycobacterium tuberculosis*, and do not require large numbers of infectious particles for efficient transmission.

Attachment

Many organisms have specific factors that allow attachment to mucosal surfaces, e.g. the cell surface lipoteichoic acid of group A β-haemolytic streptococci and pili of *N. gonorrhoeae*.

Replication and combating host defences

• The haemolysins and leukocidins of streptococci and staphylococci reduce function (chemotaxis, phagocytosis) and viability of phagocytic cells.

• Surface components of some pathogenic bacteria inhibit the activation and deposition of complement, preventing opsonization and phagocytosis by neutrophils, e.g. the capsule of *S. pneumoniae*.

• Organisms have a number of strategies for neutralizing the effects of antibody. These include: production of soluble antigen, which neutralizes the antibody in the fluid phase and so prevents binding to the microbial surface; binding of antibody in an inverted fashion, e.g. staphylococcal Fc receptor; direct destruction of antibodies by proteases, e.g. IgA protease of *N. meningitidis*; variation in antigenic structure requiring the host to synthesize new specific antibody, e.g. antigenic drift of influenza viruses.

• Intracellular survival: some microorganisms can survive and multiply within phagocytic cells. They avoid or neutralize the oxygen-dependent killing mechanism of phagocytic cells or prevent lysosomal fusion (e.g. mycobacteria, brucellae, legionellae, listerias, and many viruses).

Toxigenicity

Organisms can produce a variety of toxins, of which there are two main types: exotoxins and lipopolysaccharides (formerly known as endotoxins); their main properties are shown in Table 23.1.

Lipopolysaccharides ('Endotoxins')

Endotoxins (lipopolysaccharides [LPSs]) are important in the pathogenesis of Gram-negative septic

Table 23.1 Properties of exotoxins and endotoxins.

EXOTOXINS AND LIPOPOLYSACCHARIDES ('ENDOTOXINS')	
Exotoxins	**Lipopolysaccharides (endotoxins)**
Mainly produced by Gram-positive bacteria	Lipopolysaccharides—part of Gram-negative cell wall
Heat-labile proteins	Heat-stable proteins
High potency	Liberated when Gram-negative bacteria lyse
Strong antigenicity	Non-specific effect
Neutralized by antitoxin	
Often possess specific mechanisms of action	

shock (endotoxic shock). The LPS molecule activates the complement and cytokine pathways, releasing activated components that increase vascular permeability and result in shock; the clotting and fibrinolytic cascades are also activated, resulting in haemorrhage and thrombosis (disseminated intravascular coagulopathy).

Exotoxins

Some exotoxins provide obvious benefits to the organism in protecting against host defences or promoting transmission. The following are examples of exotoxins in disease:

• *Diphtheria toxin* produced by *Corynebacterium diphtheriae* inhibits protein synthesis and causes necrosis of the epithelium, heart muscle, kidney and nerve tissues.

• *Tetanus toxin* produced by *Clostridium tetani* increases reflex excitability in neurons of the spinal cord by blocking release of inhibitory neurotransmitters (glycine and γ-aminobutyric acid) in motor neuron synapses.

• *Botulinum toxin* produced by *Cl. botulinum* affects motor neurons, blocking release of acetylcholine at synapses and neuromuscular junctions and resulting in motor paralyses (e.g. dysphagia, respiratory arrest).

• *Staphylococcal enterotoxin* produced by specific strains of *S. aureus* acts on the CNS, resulting in severe vomiting within hours.

• *Clostridial toxins* produced by *Cl. perfringens* cause tissue necrosis and haemolysis.

• *Cholera toxin* produced by *Vibrio cholerae* binds to ganglioside receptors of the small intestine epithelial cells and results in increased adenylyl cyclase activity, with massive hypersecretion of chloride and water into the lumen.

Host responses to infection

The human body protects itself against infection by several mechanisms:

• *Non-specific or innate immune defences* represent the initial defences encountered by an organism when it comes into contact with the host.

• *Specific or adaptive mechanisms* are designed to act when non-specific defences are breached; these involve recognition by the body of specific components of the invading organism (antigens) and the production of immune factors (antibodies, T lymphocytes); finally it commits the antigenic detail to immunological memory.

On first encountering a pathogen the specific immune system is slow to react and the body is heavily reliant on non-specific immune defences. However, subsequent exposure to the same organism elicits a much more rapid response ('booster' response) from the specific immune system; the host has acquired immunity.

The immunology of infectious disease is complex and, to help understand it, it is necessary to describe the different systems involved in host defence as separate entities ('non-specific' and 'specific'; 'antibody-dependent immunity' and 'cell-mediated immunity'). However, these components are interdependent and often act synergistically, being coordinated by cytokines.

Table 23.2 The importance of physical defences.

IMPORTANCE OF PHYSICAL DEFENCES			
Site of entry	Defence mechanism	Condition of defence reduction	Consequence
Skin	Dry and acidic (fatty acids); continuous shedding of normal flora	Burns, wounds, eczema, ulcers	Colonization and infection by *Staphylococcus aureus* and *Pseudomonas* species
Alimentary tract	Acid in stomach	Achlorhydria	Oral candidiasis
	Motility	Antibiotic treatment	Overgrowth of coliforms in upper gastrointestinal tract, resulting in vomiting and in aspiration pneumonia
	Mucus coating	Ileus (post-surgery)	
	Normal flora		
	IgA antibodies		
Respiratory tract	Nasal filters	Cystic fibrosis	Chronic lung infection
	Mucus trapping	Smoking	
	Ciliary action	Post-surgery	Postoperative pneumonia
	Coughing	Intubation	
	IgA antibodies		
Urinary tract	Outward flow	Vesicoureteric reflux	Recurrent and chronic urinary tract infection
	Complete emptying bladder	Outflow obstruction with residual urine in bladder	
	Mucus coating		
	IgA antibodies		
Genital tract (female)	Acid production by lactobacilli (normal flora)	Pregnancy and oral contraceptives (reduced acidity)	Vaginal candidiasis
	IgA antibodies	Antibiotic therapy	

Non-specific immune defences

Non-specific defences can be divided into:
- external biochemical and mechanical barriers
- phagocytic cells
- soluble factors (e.g. complement)
- natural killer (NK) cells.

External biochemical and mechanical barriers

The importance of these external barriers is best illustrated by considering the infections that result from their reduction (Table 23.2).

Phagocytes

The principal phagocytic cells of the body are macrophages and polymorphonuclear leukocytes (polymorphs or neutrophils).

Macrophages

Macrophages are part of the mononuclear phagocytic cell lineage. Promonocytes in the bone marrow mature into macrophages, and migrate to various body tissues, particularly lymph nodes and spleen, and submucosal tissues. They are often given specific names relating to each tissue (e.g. liver, Kupffer's cells; lung, alveolar macrophages) and have various functions including phagocytosis and killing of organisms, and the synthesis of complement components and cytokines (see below).

Polymorphs

Polymorphs originate from precursor cells in the bone marrow, and are found primarily in the circulation, only appearing in tissues as part of an inflammatory response. The majority of circulating polymorphs are sequestrated in the spleen and capillary beds, and are recruited in response to infection. The appearance of polymorphs in the blood results in the increase in peripheral white cell count associated with acute inflammation, such as that cause by bacterial infection. The primary function of polymorphs is phagocytosis and intracellular killing of microorganisms.

Phagocytosis and intracellular killing

Phagocytosis is the process whereby phagocytic cells engulf particulate matter such as bacteria; it involves several stages (Figure 23.1).

Chemotaxis
Phagocytes are attracted to the site of inflammation (chemotaxis) in response to a variety of soluble stimuli (chemotaxins), including bacterial products and inflammatory mediators produced by the host (e.g. the complement components C3a and C5a—see below).

Attachment
Bacteria attach to the phagocytic cell membrane by non-specific mechanisms dependent on the physiochemical properties of the bacterial cell surface.

Engulfment
Bacteria are engulfed by the phagocytic cells in a manner similar to that of amoebae engulfing particles, with small pseudopodia developing around the bacterial cell that eventually coalesce and internalize the organism in a 'phagosome'.

Intracellular killing
There are two principal mechanisms whereby phagocytic cells kill internalized bacteria:

1 *Oxygen-dependent killing*: toxic oxygen radicals (hydrogen peroxide, superoxide ions) produced by the phagocyte cell membrane are secreted into the phagosome; together with the lysosomal enzyme, myeloperoxidase, and halide ions, they attack the bacterial cell wall.
2 *Lysosomal killing*: the phagosome fuses with lysosomes containing enzymes that degrade the engulfed organism.

The complement system

The complement system consists of a series of serum proteins that activate each other in a cascade, similar to the clotting system (Figure 23.2).

Complement activation

The principal activator is antibody bound to antigen; other activators include bacterial lipopolysaccharide (Gram-negative bacteria) and lipoteichoic acid (Gram-positive bacteria).

Complement pathways

There are two complement pathways: the classic pathway and the alternative pathway; both generate immunologically active proteins (see Figure 23.2).

- C3a and C5a increase vascular permeability and act as chemotaxins, attracting phagocytes to the site of infection.
- C3b binds to the surface of bacteria and mediates immune adherence of bacteria to the phagocyte via the C3b receptor ('opsonization').
- C7/C8/C9 bind to the surface of the bacterial cell resulting in lysis, particularly of Gram-negative organisms.

Control of complement

As with any cascade system, feedback loops and specific degrading enzymes are important in controlling the production and removal of immunoactive complement components.

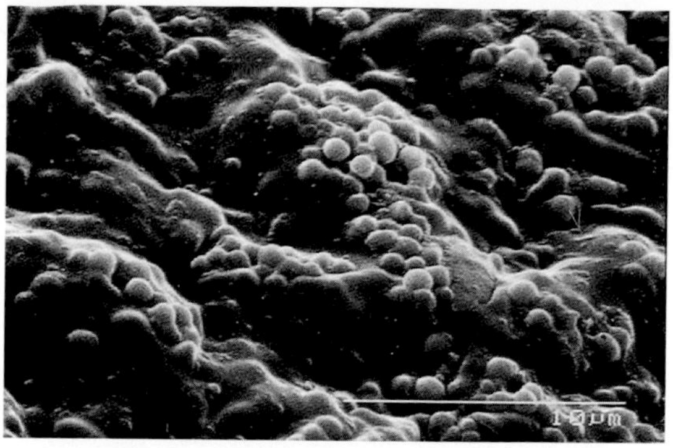

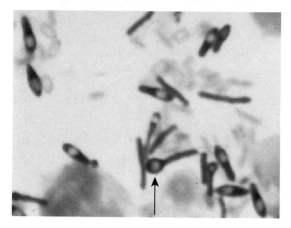

Plate 1 Scanning electronmicrograph of *Staphylococcus epidermidis* embedded in slime attached to a catheter.

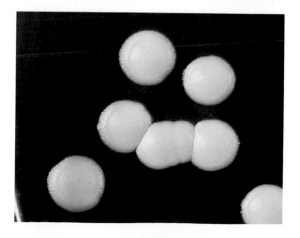

Plate 2 Gram-stain of *Clostridium sporogenes* (showing oval subterminal spores) and a *Clostridium tetani* with a terminal spore (arrowed).

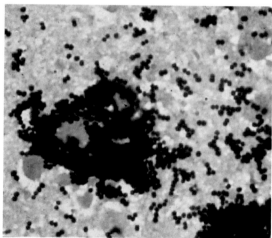

Plate 3 Gram-stained sputum obtained from patient with *Staphylococcus aureus* pneumonia. Characteristic clusters of Gram-positive cocci (1 μm diameter) stained dark purple with 'grape-like' appearance.

Plate 4 *Staphylococcus aureus* colonies on a blood agar plate (2–3 mm diameter).

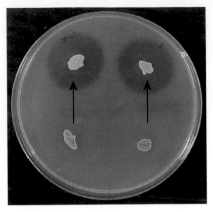

Plate 5 Plate containing DNA showing clear zones around DNAase-producing staphylococci (arrowed). DNAase-negative staphylococci shown below.

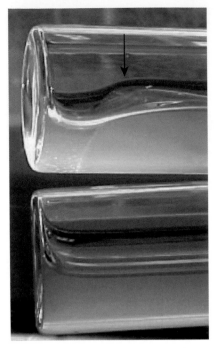

Plate 6 (upper) Fibrin clot (arrowed) produced by coagulase-positive staphylococci. (lower) Absence of fibrin clot in presence of coagulase-negative staphylococci.

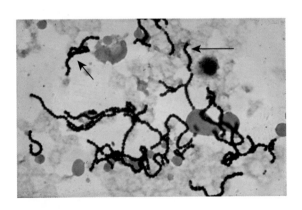

Plate 7 Gram stain of β-haemolytic streptococci showing long chains (arrowed).

Plate 8 Pale green colonies of α-haemolytic streptococci on a blood agar plate resistant to bacitracin sensitivity disc (arrowed).

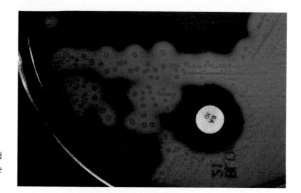

Plate 9 β-haemolytic streptococci colonies on a blood agar plate sensitive to bacitracin (arrowed). Note the zone of inhibition around the bacitracin disc.

Plate 10 *Streptococcus pneumoniae* colonies (arrowed) on blood agar with characteristic 'draughtsmen'-like appearance (1 mm diameter).

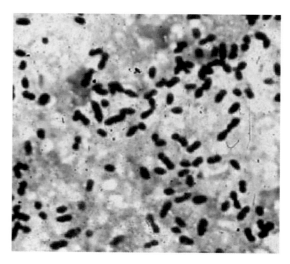

Plate 11 Gram positive cocci (1 μm diameter) of *Streptococcus pneumoniae* in pairs (diplococci).

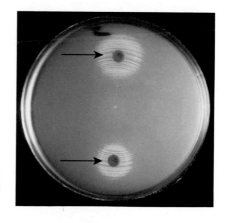

Plate 12 Agar plate with vancomycin impregnated discs showing growth of voncomycin-dependent enterococci around the discs (arrowed).

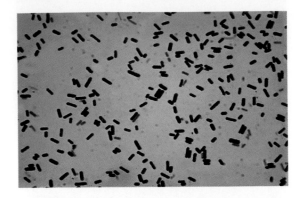

Plate 13 Gram stain of *Clostridium perfringens* showing Gram-positive bacilli with square ends (5 μm × 1 μm).

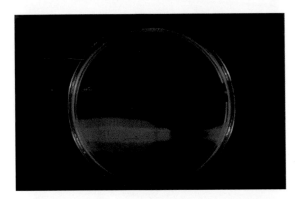

Plate 14 Nagler reaction showing lecithinase production by *Clostridium perfringens* (below) and non-lecithinase production by *Clostridium butyricum* (above).

Plate 15 Colonies of *Clostridium difficile* on a blood agar plate (2–4 mm diameter).

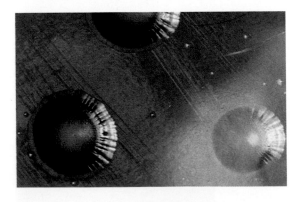

Plate 16 Black colonies of *Corynebacterium diphtheriae* with a daisy head appearance on tellurite agar (2–3 mm diameter).

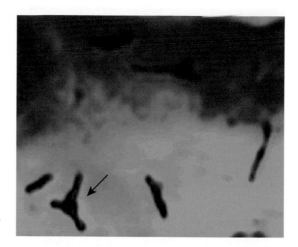

Plate 17 Gram-positive branching rods of *Propionobacterium acnes* (arrowed).

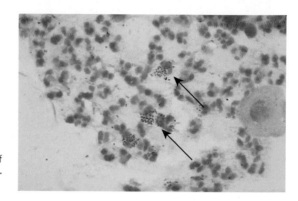

Plate 18 Intracellular Gram-negative diplococci of *Neisseria gonorrhoeae* (0.8 μm diameter) in neutrophils (arrowed).

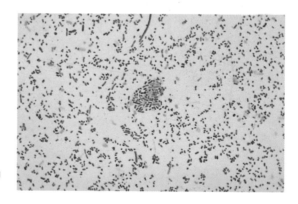

Plate 19 Gram stain of *Escherichia coli* showing Gram-negative bacilli (2 μm × 0.5 μm).

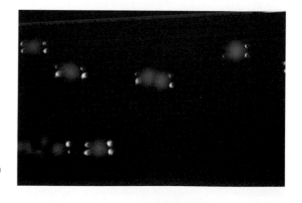

Plate 20 *Escheria coli* colonies (2–3 mm diameter) on blood agar (arrowed).

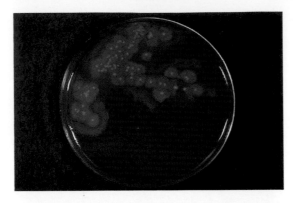

Plate 21 Colonies of Proteus species on a blood agar plate demonstrating spreading growth (swarming) (arrowed).

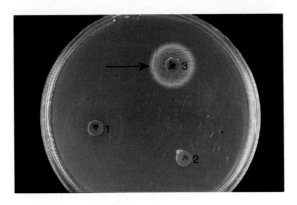

Plate 22 Nutrient agar plate with discs containing: (1) Haemin; (2) NAD; (3) both. *Haemophilus influenzae* requires both factors for growth (arrowed).

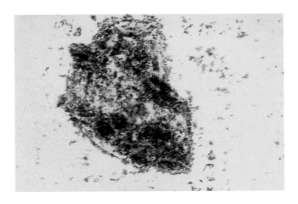

Plate 23 Microscopy of vaginal discharge, showing a single epithelial cell (clue cell) with numerous organisms attached.

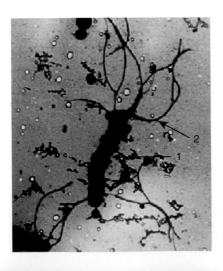

Plate 24 Negative-stain electronmicrograph of a single *Helicobacter pylori* (1) showing flagella (2).

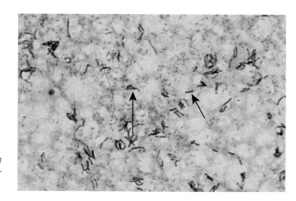

Plate 25 Black colonies of *Prevotella melaninogenica* on a blood agar plate (1–2 mm diameter).

Plate 26 Ziehl–Neelsen stain of *Mycobacterium tuberculosis* (stained red, arrowed). Acid-fast slender rods, 3 μm × 0.3 μm).

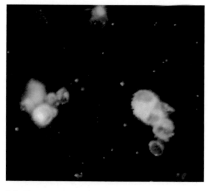

Plate 28 Fluorescence microscopy showing intracellular inclusion bodies of chlamydia.

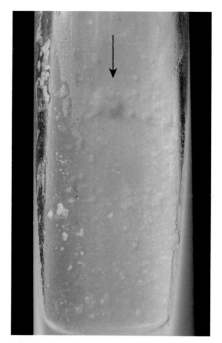

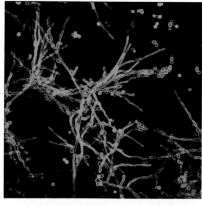

Plate 27 Colonies of *Mycobacterium tuberculosis* on Lowenstein–Jensen slopes (arrowed).

Plate 29 *Candida albicans*: both filamentous and unicellular forms stained with a fluorescent dye.

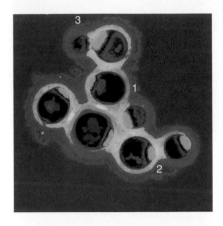

Plate 30 (1) *Cryptococcus neoformans* (capsulated strain): (2) cells stained with a fluorescent dye to demonstrate cell division and (3) daughter cell separation.

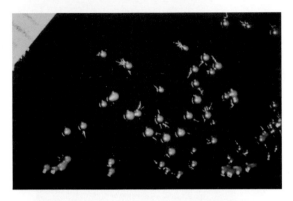

Plate 31 Colonies of *Candida albicans* on a blood agar plate (1–2 mm diameter).

Plate 32 *Aspergillus fumigatus* (left) on Sabouraud's agar. Strains produce green colonies with white periphery. *Aspergillus niger* (right) demonstrating dense black colonies with white periphery.

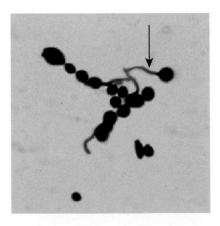

Plate 33 Gram stain of Candida albicans. Gram-positive oval budding yeast cells (6 μm × 4 μm) with germ tube production (arrowed).

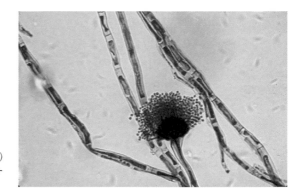

Plate 34 Microscopy of fruiting body (conidiophore) of *Aspergillus fumigatus* stained with lactophenol cotton blue.

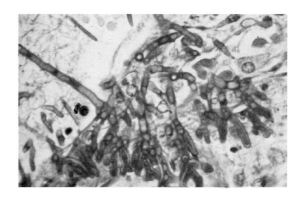

Plate 35 Invasive aspergillus hyphae in brain tissue.

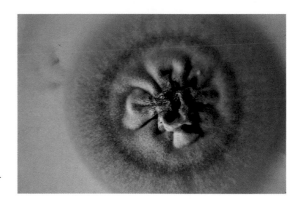

Plate 36 Colony of *Epidermophyton floccosum* illustrating its undulating surface.

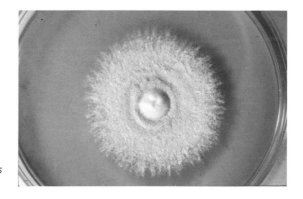

Plate 37 Colony of *Trichophyton mentagraphytes* illustrating its powdery white surface.

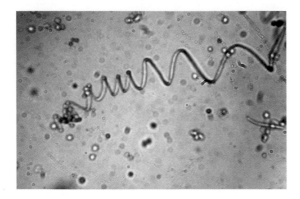

Plate 38 Microscopy of spiral hyphae of *Trichophyton mentagrophytes*.

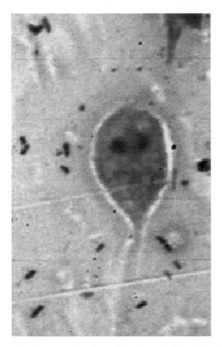

Plate 40 Giemsa-stained trophozoites of *Giardia lamblia* with its distinctive 'teardrop' shape and laterally placed nuclei.

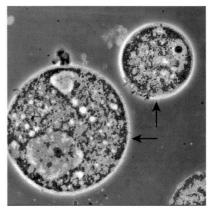

Plate 39 *Acanthamoebae* species: 'rounded' tropozoite (arrowed) stained with a fluorescent dye.

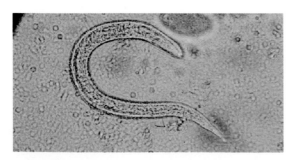

Plate 41 Microscopy of *Strongyloides stercoralis* larva in iodine preparation.

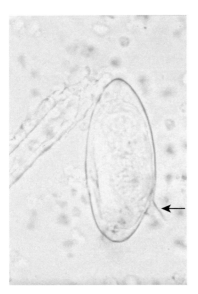

Plate 42 Ova of *Schistosoma mansoni* illustrating a smooth, thin wall and lateral spine (arrowed).

Plate 43 Class 1 safety cabinet for dealing with dangerous pathogens.

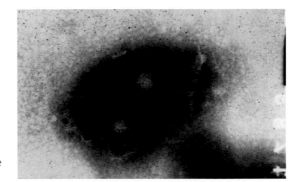

Plate 44 Herpes simplex virus seen after negative staining.

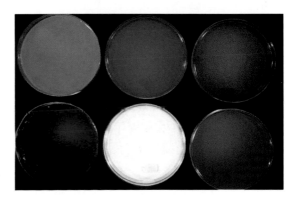

Plate 45 Some different agar plates used for identifying bacteria.

Plate 46 Anaerobic cabinet used for growing anaerobic bacteria.

Plate 48 Gonochek II for identifying *Neisseria gonorrhoeae* by biochemical reactions.

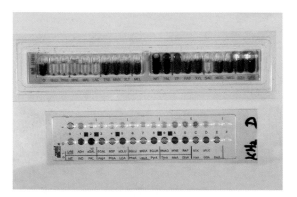

Plate 49 Commercial biochemical-based identification system (AP1).

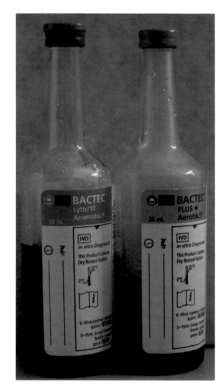

Plate 47 Aerobic and anaerobic Bactec blood culture bottles.

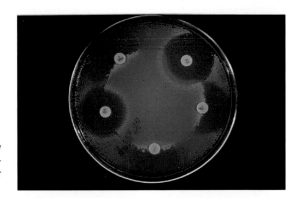

Plate 50 Antibiotic sensitivities of a staphylococcus by Stokes method. Zones of inhibition around the antibiotic impregnated discs for the test organism (inner circle) and control (outer circle) are compared.

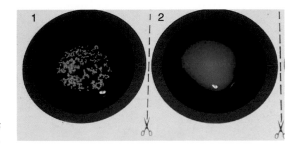

Plate 51 Latex agglutination test for *Staphylococcus aureus*: (1) positive agglutination; (2) no agglutination.

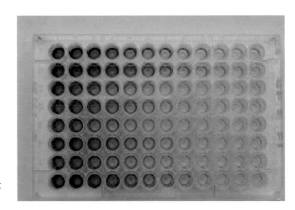

Plate 52 ELISA (enzyme-linked immunosorbent assay) plate. A frequently used serological assay.

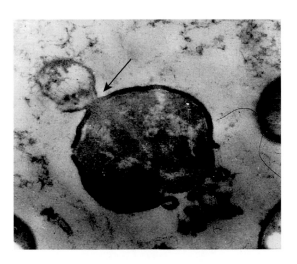

Plate 53 Electronmicrograph illustrating effect of penicillin breaking cell wall of a susceptible bacterium (arrowed).

Plate 54 Vitek® machine used for automated sensitivity testing.

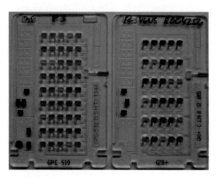

Plate 55 Vitek® cards used for automated antibiotic sensitivity testing and identification of bacterial isolates.

Plate 56 Disc diffusion sensitivity test using British Society for Antimicobial Chemotherapy (BSAC) method.

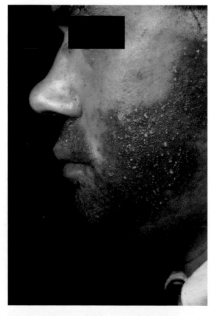

Plate 57 Staphylococcal infection of the face.

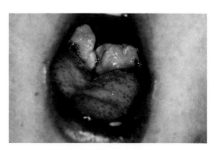

Plate 58 Case of diphtheria showing false membrane on tonsils.

Plate 59 Gonococcal urethral discharge.

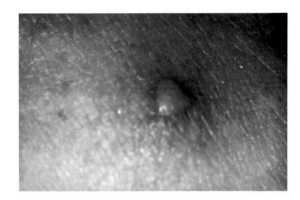

Plate 60 Gonococcal skin lesion.

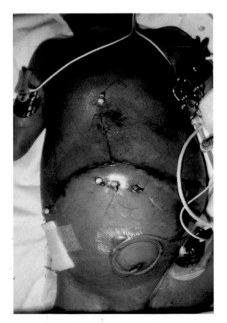

Plate 61 Liver transplant patient with numerous intravascular and drainage catheters.

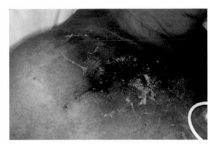

Plate 62 Catheter-related sepsis at insertion site of device.

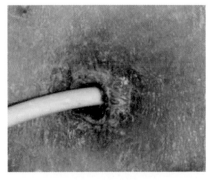

Plate 63 Localized exit site infection showing erythema and exudate.

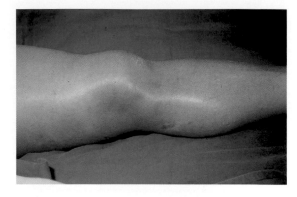

Plate 64 Septic arthritis of knee.

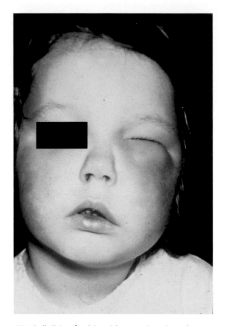

Plate 65 Cellulitis of orbit with associated erythema.

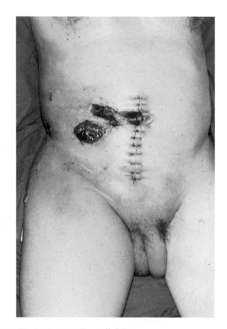

Plate 66 Post-operative cellulitis.

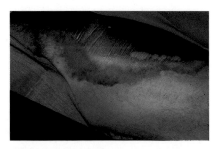

Plate 67 Gas gangrene of leg.

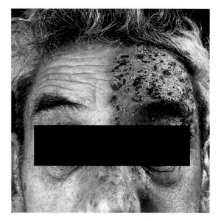

Plate 68 Shingles on face.

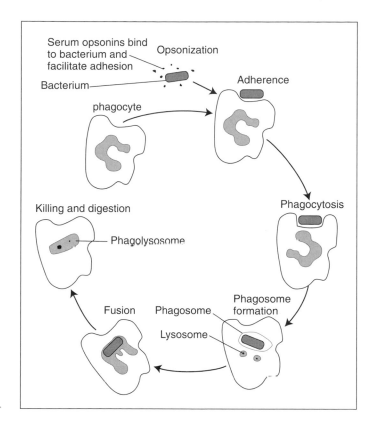

Figure 23.1 The phagocytic process.

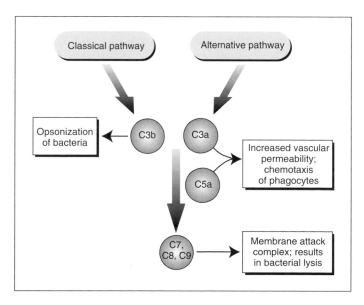

Figure 23.2 The complement pathway.

Complement and disease

The absence of complement components or enzymes involved in feedback mechanisms can lead to a spectrum of disease ranging from recurrent infection to immunopathological conditions such as systemic lupus. Deficiencies in C7/C8/C9 are associated with recurrent meningococcal infection.

Natural killer cells

NK cells are large granular lymphocytes that provide a non-specific response to virally infected cells and tumour cells. The NK cells bind to viral or tumour surface proteins on the abnormal host cells, and deliver the cytotoxic contents of their granules.

Specific defence mechanisms

Virulence factors such as capsules and toxins are used to avoid the non-specific immune defences or to produce disease. In response, the host has developed a specific immune system that is based principally on the recognition of foreign proteins, carbohydrates and other molecules (antigens) by specific immunoglobulin (antibodies) and cells (T lymphocytes). These neutralize the invading organism or its products (e.g. toxins) and crucially support the non-specific immune system to interfere with the pathogen.

Antigens

The term 'antigen' describes particular microbial molecules to which specific antibodies bind, the system being comparable to a lock and key. Antigens on bacterial cell surfaces are often carbohydrates (cell wall and capsular antigens). Protein antigens include the nucleoproteins and glycoproteins of viruses and bacterial exotoxins or surface proteins.

Antibodies

Human antibodies (immunoglobulins) can be classified into five major classes: IgG, IgA, IgM, IgD and IgE; they differ in size and biological functions (Table 23.3). All have the same basic unit structure (Figure 23.3).

The Fab (fragment antigen-binding) portion of the antibody is the recognition site that binds to complementary antigen. The Fc site of the antibody binds to Fc receptors present on phagocytes, facilitating phagocytosis (opsonization). The Fc receptor can also activate complement by binding to the C3b complement component. Antibodies may also prevent viral attachment and therefore

Table 23.3 Antibody classes: composition and function.

ANTIBODY CLASSES			
Antibody class	Approximate molecular weight	Composition	Comments
IgG	160 000	1 polypeptide unit	80% of total immunoglobulin four major subclasses (1–4) Crosses placenta
IgM	900 000	5 polypeptide units	First antibody to appear following infection Does not cross placenta
IgA	160 000–320 000	1 or 2 polypeptide units	Principal mucosal antibody; important in host defence at mucosal surfaces Present in breast milk
IgE	200 000	1 polypeptide unit	Key role in hypersensitivity reactions (allergy) Important in defence against parasitic infections
IgD	200 000	1 polypeptide unit	Mainly found on surface of B lymphocytes

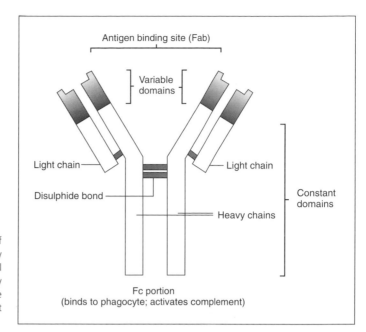

Figure 23.3 Basic unit structure of immunoglobulin. Two identical heavy polypeptide chains and two identical light polypeptide chains are linked by disulphide bonds. The domains at the Fab portion are variable to reflect binding to different antigens.

invasion of host cells. Antitoxin antibodies bind to bacterial exotoxins and prevent their biological action.

Lymphocytes

Lymphocytes are derived from stem cells that originate in thymus and bone marrow. They are found in lymph nodes and spleen. Lymphocytes are primarily divided into *B cells* (so-called because they mature in the bone marrow) and *T cells* (which mature in the thymus). Lymphocytes can be further subdivided according to molecules on their surface; these molecules are known as cluster determinants (CD). CD surface markers are important in the subdivision of T cells.

B lymphocytes

B lymphocytes express immunoglobulins on their surfaces. During differentiation each B lymphocyte becomes committed to the expression of an Ig specific for a particular antigen; the B cell is then said to be immunocompetent. Foreign antigen (e.g. an invading bacterium) is recognized by lymphocytes expressing the complementary antibody; this leads to clonal proliferation of the original cell, resulting in a large population of B cells secreting the correct antibody. B cells actively secreting antibody undergo morphological changes and are termed 'plasma cells'.

The clonal proliferation of B lymphocytes after initial contact with antigen results in not only the production of large numbers of active plasma cells but also a larger pool of B lymphocytes that express the particular Ig. These cells are known as memory cells and are responsible for the more rapid antibody response that occurs when the host is challenged with the same antigen a second time (Figure 23.4). After a second stimulation there is normally a very low, or no, IgM response. Such primary and secondary antibody responses are important when considering immunization (see Chapter 27).

T lymphocytes

T lymphocytes also have surface receptors that recognize antigens. As with B cells, T cells express a single receptor specific for a particular antigen and clonal expansion of individual T cells occurs in response to an antigen. Subtypes of T cells with

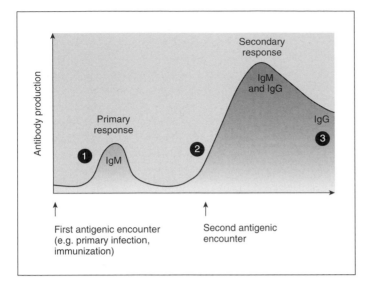

Figure 23.4 Kinetics of antibody response. (1) First contact with an antigen requires recognition and clonal proliferation of specific B lymphocytes before antibody production (principally IgM). (2) Subsequent contact with the antigen results in more rapid antibody production (IgM and IgG). (3) Serum IgG antibodies remain raised and can be measured to confirm immune status.

Table 23.4 T-lymphocyte subtypes and functions.

T-LYMPHOCYTE SUBTYPES AND FUNCTIONS	
Subtype	**Function**
CD4 (T-helper cells)	Stimulate B lymphocytes to synthesize antibody
	Cytokine production
CD8 (T-suppressor cells)	Inhibit T-helper and B lymphocytes
	Control of immune response
CD21/CD35	Complement synthesis
CD8 (cytotoxic T cells)	Recognition of microbial antigens in complexes with cell surface

different functions are recognized by CD surface markers (Table 23.4).

T lymphocytes and macrophages are important in cell-mediated immunity against viruses and bacteria (e.g. mycobacteria) that can survive phagocyte and antibody/phagocyte defences. Infected phagocytes are recognized and destroyed by recruitment (via cytokines) of macrophages and cytotoxic T lymphocytes.

Cytokines

Cytokines are soluble factors involved in intercellular communication and include interleukins (various types), interferons (α, β, γ), tumour necrosis factor (TNF) and the colony-stimulating factors (e.g. granulocyte–macrophage colony-stimulating factor [GM-CSF] which stimulates growth of granulocytes and monocytes).

The cytokines are the subject of intensive investigation, and a full understanding of the structure and function of individual factors is not yet elucidated. Functions include the proliferation and maturation of T and B cells, the activation of polymorphs, antiviral effects, cytotoxicity, and stimulation of growth of granulocytes and monocytes.

Research into cytokines has not only improved our understanding of the immune response, but also led to the development of novel therapeutic agents (e.g. GM-CSF for the treatment of neutropenia; monoclonal antibody to TNF for the treatment of endotoxic shock; interferons as antiviral agents).

Disease secondary to immunopathology

Over-reaction of the immune response ('hypersensitivity') may result in tissue damage that results in clinical disease. The disease is categorized into five types:

1. Type I, resulting from release of IgE from mast cells, e.g. release of parasite antigens when a hydatid cyst ruptures.

2. Type II, mediated by IgG to surface complement and cytotoxic cells, e.g. virus-infected cells.

3. Type III, caused by deposition of immune complexes of antigen-antibody, complement and polymorphs, e.g. glomerulonephritis secondary to group A streptococcal infection.

4. Type IV: tissue destruction by T lymphocytes, cytokines and macrophages, e.g. the granulomata of mycobacterial infection (reactivated TB, tuberculoid leprosy),

5. Autoimmunity: antibodies that cross-react with host tissues, e.g. myocarditis secondary to group A streptococcal infection.

Activation of complement, inflammatory mediators, and clotting by lipopolysaccharide (or lipoteichoic acid) can cause septic shock and associated disseminated intravascular coagulation.

Chapter 24

Diagnostic laboratory methods

This chapter describes the common techniques used in diagnostic microbiology laboratories.

General principles for specimen collection

1 Whenever possible, specimens should be obtained before starting or altering antimicrobial therapy, and be collected into sterile containers by techniques that avoid contamination from the normal flora of the patient, the person obtaining specimens or the environment.

2 Delays in transportation to the laboratory may reduce viability of some organisms and result in overgrowth of others. Special transport media are available but rapid transport to the laboratory is preferable, particularly for fastidious organisms, e.g. *Neisseria gonorrhoeae*.

3 Specimens should be transported in leak-proof containers in accordance with safety guidelines.

4 The request form should document the patient's basic clinical details, including age, diagnosis, date of onset of symptoms, date and time of specimen collection, and current antimicrobial therapy. When appropriate, history of travel, suspected contact of infection, immunization or involvement in an outbreak should be given. This information is essential for the laboratory in selecting which investigations to perform and for interpreting results.

Laboratory diagnosis of infection

Samples can be examined by microscopy, culture and various antigen–antibody tests. High-risk specimens, possibly containing category 3 organisms, e.g. sputum, from a suspected case of tuberculosis (TB) are prepared in special cabinets (Plate 43).

Direct microscopy

Samples can be examined either directly (e.g. urine) or in an emulsified suspension (e.g. faeces). Direct microscopy is normally limited to the following investigations:

• The detection of parasites (helminths, protozoa) by visualization of trophozoites, ova or cysts (see Plates 40 and 41): faeces for intestinal parasites; urine for schistosomes (see Plate 42); vaginal swabs for *Trichomonas*; blood for malaria.

• Direct microscopy of urine and cerebrospinal fluid (CSF) samples in order to assess the type and number of inflammatory cells.

Direct microscopy following staining

Gram stain

The Gram stain is the principal stain used in diagnostic microbiology and allows rapid classification of bacteria into four simple categories: Gram-positive cocci, Gram-positive bacilli, Gram-negative cocci or Gram-negative bacilli (see Plates

2, 3, 7, 11, 13, 17, 18 and 19). Fungi can also be identified by the Gram stain. It also helps to determine the presence of inflammatory cells. In most diagnostic microbiology laboratories, Gram stains are performed routinely on specimens of sputum, pus, CSF and swabs of the genital tract. These are samples where the presence and/or morphology of commensals will not be confused with infecting organisms seen on the Gram stain.

GRAM STAIN

1 Spread specimen thinly on glass microscope slide and heat fix
2 Stain with crystal violet (blue)
3 Add Lugols iodine (this fixes the dye in Gram-positive bacteria only)
4 Decolourize with acetone or alcohol (only Gram-negative bacteria are decolourized; Gram-positive remain blue)
5 Counter stain; carbol fuchsin (stains Gram-negative bacteria red; Gram-positive bacteria remain blue/purple)

Ziehl–Neelsen (ZN) stain

ZN stains are performed on specimens, including sputum, pus and urine when mycobacterial infection is suspected (see Plate 26).

ZIEHL–NEELSEN STAIN

1 Spread specimen thinly on glass microscope slide and heat fix
2 Stain with **HOT** carbol–fuchsin dye (all bacteria are stained)
3 Decolourize with 20% sulphuric acid (only mycobacteria retain dye)
4 Counter stain with methylene blue or malachite green to reveal non-mycobacterial organisms

---

Other special stains

A number of other stains are used occasionally In the clinical laboratory; these are referred to in the sections describing specific pathogens.

Electron microscopy

Electron microscopy is used to visualize and ident-

ify viruses in some specimens, e.g. faecal samples from children with diarrhoea (i.e. rotaviruses) or from patients involved in outbreaks of food poisoning (i.e. noroviruses); or blister fluids for the rapid diagnosis of herpes simplex (Plate 44) and varicella-zoster infections.

Culture techniques

- Most medically important bacteria and fungi can be cultured on laboratory media (see Plates 4, 8–10, 15, 16, 20, 21 and 25). Culture remains the most reliable method for confirming diagnosis and optimizing treatment by the performance of antibiotic susceptibility tests.
- Various media (Plate 45) and culture conditions are utilized for different specimens, reflecting the likely pathogens. Culture plates are usually incubated either under aerobic conditions (sometimes with added CO_2) or in an anaerobic atmosphere in specialized cabinets (Plate 46).
- There are two basic types of specimen:
  (1) *specimens from normally sterile sites* (e.g. blood [Plate 47], CSF, joint fluid): cultured on enriched media to isolate most organisms because any isolate is likely to be significant
  (2) *specimens from sites with a normal flora* (e.g. upper respiratory tract specimens, faeces, genital specimens): cultured on selective media designed to suppress normal endogenous flora but allow likely pathogens to grow.
- Sputum and urine specimens are sometimes described as 'clean contaminated', because normally both should be sterile, although they may become contaminated during sampling: sputum by upper respiratory tract flora or urine by perineal flora. Some 'contaminating organisms' may also cause infection (e.g. pneumococci from the pharynx may contaminate sputum but are also an important lower respiratory tract pathogen). Quantitative culture is performed to help distinguish contamination (low numbers of bacteria/several different bacterial species) from infection (high numbers of single bacterial species). This approach is particularly important in the diagnosis of urinary tract infection (Chapter 36).

Virus culture and identification methods are described in Chapter 18.

Microbial identification

Techniques to identify bacteria and fungi include:

- growth characteristics (e.g. aerobic or anaerobic)
- colonial morphology (e.g. size, haemolysis, colour)
- microscopy of organism with or without staining (wet preparations of fungal colonies are particularly important in speciation of medically important fungi)
- antigenic characteristics (e.g. Lancefield grouping of β-haemolytic streptococci; speciation of *Salmonella*)
- biochemical tests (e.g. fermentation of sugars, detection of specific enzymes; Plates 48 and 49)
- growth factor requirements (e.g. X and V dependency of *Haemophilus influenzae*—see Plate 22)
- production of toxins (e.g. Nagler test for *Clostridium perfringens* (see Plate 14 and Chapter 7); Elek test for *Corynebacterium diphtheriae* (Figure 24.1; see Chapter 8).

Identification to species level is normally satisfactory in a diagnostic laboratory. Further characterization of individual isolates ('typing') may be important when defining outbreaks or cross-infection; techniques include phage typing, plasmid analysis and characterization of chromosomal DNA.

Antibiotic sensitivity testing

Significant isolates are tested for susceptibility to different antimicrobial agents by various techniques (e.g. disc testing—Plates 50 & 56); these are described in Chapter 29.

Immunological methods

Antigen–antibody interactions have developed into an important part of laboratory diagnosis. The basic principle is that for each microbial antigen there is a specific antibody. Antigen–antibody reactions can be used in developing tests to detect *specific organisms* or a *specific antibody*. The antigen–antibody reaction created must be made visible and can be achieved by various techniques.

Precipitation reactions
These include the Elek test for diphtheria toxin (see Figure 24.1).

Agglutination reactions
An example of this is speciation of *Salmonella*; known antisera to the various *Salmonella* species are mixed individually with suspensions of an unknown salmonella isolate; an agglutination reaction (visible clumping of the bacteria) with a particular antisera indicates the serotype.

Co-agglutination reactions
Known antibody *or* antigen is fixed to an inert particle (e.g. latex beads). When an antigen–antibody reaction occurs, the beads agglutinate; examples include:

- Lancefield grouping of streptococci
- rapid diagnosis of bacterial meningitis by detection of bacterial antigen in CSF
- identification of *Neisseria gonorrhoeae*
- identification of *S. aureus* (Plate 51)
- detection of rotaviruses in stool specimens.

Antigen detection (Table 24.1)
Fluorescein-labelled specific antibodies may be used to detect some pathogens directly in specimens (Figure 24.2); the specimen is viewed by fluorescence microscopy. In positive samples fluorescent foci are seen (Figure 24.2 and Plates 28, 29, 30 & 31). The techniques are used in routine diagnostic microbiology for the rapid diagnosis of respiratory viral infections (see Plate 28) and for serious infections (e.g. diagnosis of bacterial meningitis), or where a pathogen is difficult or impossible to culture (e.g. *Pneumocystis jiroveci*).

Specific antibodies may also be labelled with enzymes (enzyme-linked immunosorbent assay [ELISA]) which, after binding, can be visualized by the addition of a substrate that is converted to a coloured product (Plate 52). The end-point may be read by a colorimeter. Examples include the

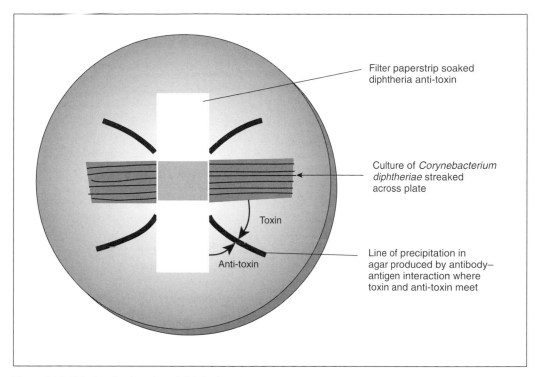

Filter paperstrip soaked diphtheria anti-toxin

Culture of *Corynebacterium diphtheriae* streaked across plate

Toxin

Anti-toxin

Line of precipitation in agar produced by antibody–antigen interaction where toxin and anti-toxin meet

Figure 24.1 Elek plate for detection of diphtheria toxin.

Table 24.1 Direct detection of organisms or their antigens in clinical specimens.

ORGANISM AND ANTIGEN DETECTION		
Specimen	**Organism**	**Technique**
Respiratory secretions	Influenza viruses	Immunofluorescence
	Parainfluenza viruses	
	Respiratory syncytial virus	
	Adenoviruses	
	Measles virus	
	Legionella pneumophila	
	Pneumocystis jiroveci	
	Group A streptococci	ELISA
Urogenital specimens	*Chlamydia trachomatis*	ELISA or molecular methods
Cerebrospinal fluid	*Haemophilus influenzae*	Co-agglutination
	Streptococcus pneumoniae	
	Neisseria meningitidis	
	Group B streptococci	
Faeces	Rotaviruses	ELISA
	Adenoviruses	Electron microscopy

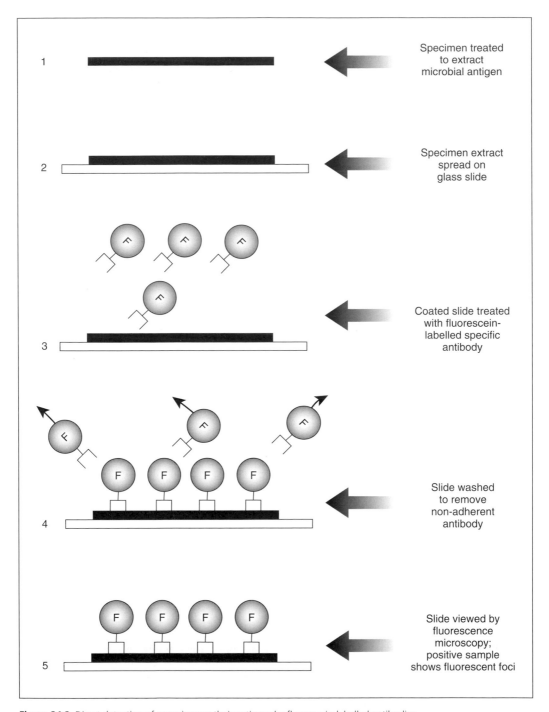

1   Specimen treated to extract microbial antigen

2   Specimen extract spread on glass slide

3   Coated slide treated with fluorescein-labelled specific antibody

4   Slide washed to remove non-adherent antibody

5   Slide viewed by fluorescence microscopy; positive sample shows fluorescent foci

Figure 24.2 Direct detection of organisms or their antigens by fluorescein-labelled antibodies.

detection of adenoviruses, rotaviruses and noro-viruses in stool samples.

Serology

Detection of antibody by labelled antibody techniques requires an additional step. These assays are often carried out in microtitration trays. Individual wells are precoated with known microbial antigen and the patient's serum added. The well is washed to remove non-specific binding and the remaining bound antibody detected with antibody to Ig-labelled with fluorescein or enzyme (ELISA—Figure 24.3, Plate 52). Antibodies of the various human antibody classes, IgG, IgM or IgA, can be raised, and each can be detected individually. Another method for capturing specific IgG and IgM antibodies is to coat wells with anti-human immunoglobulins. The patient's serum is added to the wells and bound antibodies are detected by the addition of specific antigens followed by a labelled antibody (Figure 24.3).

• Various serological techniques are used in diagnostic microbiology:

— ELISA and fluorescent labelling are described above; many commercial and automated kits are now available

— complement fixation test: assay based on the degradation of complement in the presence of antigen–antibody complexes (measured by lack of lysis of haemolysin–erythrocyte complexes, 'activated red blood cells'); gradually being replaced by ELISA

— agglutination and co-agglutination methods; often used for rapid screening of sera.

• Serology is important for the diagnosis of infections caused by pathogens that are difficult to culture by standard laboratory methods, e.g. *Mycoplasma pneumoniae*, syphilis, hepatitis A, B and C (HAV, HBV, HCV), HIV.

• Active infection is diagnosed by either detection of specific IgM or a fourfold increase in IgG antibodies in paired sera (acute and convalescent samples) taken 10–14 days apart.

• Immune status testing (e.g. to test response to hepatitis B or rubella immunization) is performed by detection of IgG antibodies.

Other diagnostic methods

Gas–liquid chromatography

This can detect the presence of volatile fatty acids produced by anaerobic bacteria in pus samples. It is occasionally used for the rapid detection of anaerobes in clinical samples or identification of some bacteria, e.g. *clostridiun difficile*.

Nucleic acid detection

Molecular biology techniques for the detection of microbial nucleic acids in specimens have been developed and are gradually being introduced into microbiology laboratories for the diagnosis of a large number of organisms. The first assays developed relied on DNA hybridization and lacked sensitivity. Newer assays that amplify nucleic acids are now available; these have much greater sensitivity and are useful for organisms that are difficult or slow to culture (e.g. mycobacteria, chlamydiae, viruses, such as HBV, HCV, HIV. Most of the currently available techniques can be used to provide quantitative results that are useful in the diagnosis of disease, staging of infection and prediction of disease, and for monitoring the response to antimicrobial therapy. A number of these assays are now available as commercial kits and in automated systems with the advantage of standardization and improved quality control. Available techniques include:

• *Polymerase chain reaction (PCR):* uses thermostable DNA polymerase to extend oligonucleotide primers complementary to the target nucleic acid in consecutive cycles of denaturing, annealing and extension. This results in an exponential accumulation of target DNA (Figure 24.4). The technique can be used as a 'multiplex PCR' to detect several targets within a specimen in one PCR reaction. *Real-time PCR* is another variation of the technique, which allows the reaction to be undertaken within a closed system using very rapid temperature cycling times and resulting in completion of a PCR reaction and detection of the product within minutes.

• *Ligase chain reaction (LCR):* involves hybridization of two oligonucleotide probes at adjacent

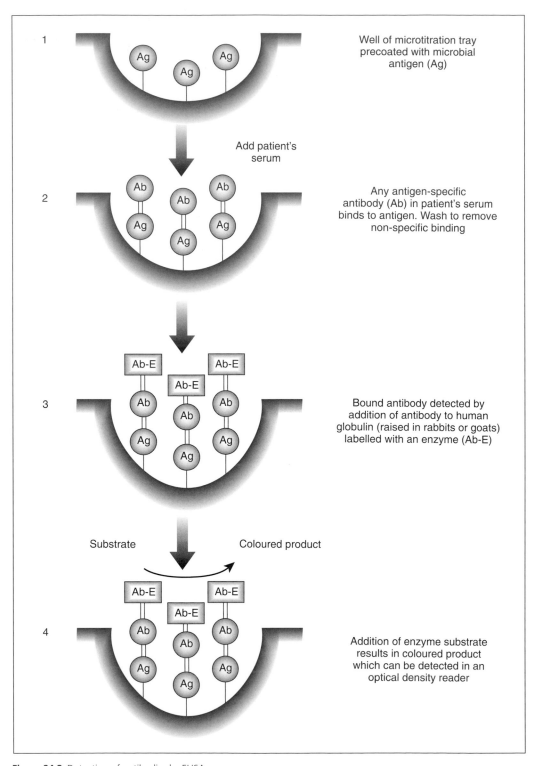

1   Well of microtitration tray precoated with microbial antigen (Ag)

Add patient's serum

2   Any antigen-specific antibody (Ab) in patient's serum binds to antigen. Wash to remove non-specific binding

3   Bound antibody detected by addition of antibody to human globulin (raised in rabbits or goats) labelled with an enzyme (Ab-E)

Substrate   Coloured product

4   Addition of enzyme substrate results in coloured product which can be detected in an optical density reader

Figure 24.3 Detection of antibodies by ELISA.

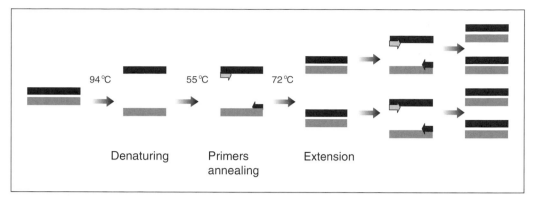

Figure 24.4 PCR is used to amplify the amount of a particular DNA molecule in a sample. Primers complementary to particular regions of the DNA of interest are added to a sample along with dNTPs and a thermostable DNA polymerase enzyme. A complementary strand of DNA is built in successive thermal cycles using the primer as the initial building block. The process can then be repeated for the resulting complementary strand leading to exponential amplification of the original DNA present in the sample.

positions on a strand of target DNA that are subsequently joined by a thermostable ligase enzyme.

- *Nucleic acid sequence-based amplification (NASBA):* uses RNA as a target and utilizes three enzymes simultaneously: reverse transcriptase, RNAase H and DNA-dependent RNA polymerase
- *Transcription-mediated amplification (TMA):* uses similar principles to NASBA
- *Strand displacement amplification (SDA):* utilizes oligonucleotide primers containing a restriction enzyme site, DNA polymerase and a restriction enzyme at a constant temperature to produce exponential amplification of the target.

- *Branched chain DNA (bDNA):* the technique uses signal amplification methods rather than amplifying target genome. It utilizes an assortment of short oligonucleotide hybrid probes that anneal to the target DNA and to each other to produce a detectable signal.

Chapter 25

Epidemiology and prevention of infection

Community-acquired infections are acquired in the community not in hospitals. Hospital-acquired ('nosocomial') infections are those where the infecting organism is acquired during hospitalization. Normally, such infections are evident during the patient's hospital stay, but, occasionally (e.g. wound infection), the infection presents after the patient's discharge home. Studies in the UK and the USA have estimated the incidence of hospital-acquired infection to be between 5 and 10% of all inpatients. The principal sites of infection are the urinary tract, wounds and lower respiratory tract, and infections associated with intravascular catheters.

Increased risk of infection is related to: prolonged hospitalization; intensive care; the use of invasive, prosthetic devices (e.g. intravenous catheters, urinary catheters, endotracheal tubes); an immunocompromised host, including those with impaired local host defences (e.g. burns, trauma). The widespread use of antibiotics in hospitals results in the selection of antibiotic-resistant microorganisms in the hospital environment; patients in hospital who receive antibiotics frequently become colonized with antibiotic-resistant microorganisms, which may subsequently cause infection.

Strategies to reduce the risk of infection in individuals or in communities are dependent on an understanding of the sources and modes of transmission of microorganisms.

Epidemiology of infection

Epidemiology is the study of the aetiology and occurrence of infections in populations. Factors affecting the epidemiology of infections are shown in Figure 25.1. Infections may be either sporadic, or occur in outbreaks, with two or more related patients, suggesting that transmission has occurred.

Definitions
• The *incidence rate* is the number of new cases of acute disease in a specified population over a defined period and is expressed as a proportion of the total population of the community involved.
• The *prevalence rate* is the total number of cases present in a defined population at a certain time point or over a certain period.
• The *attack rate* is the proportion of the total population at risk who became affected during an outbreak.
• Infections that remain present in the population are called *endemic*; an increase in incidence above the endemic level is described as hyperendemic.

Sources of infection

Endogenous
These are infections caused by microorganisms that are part of the host's normal flora.

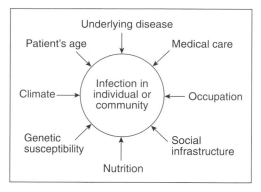

Figure 25.1 Factors affecting the epidemiology of infection.

Table 25.1 Examples of oral transmission of infectious agents

ORAL TRANSMISSION	
Vector	**Infection (example)**
Direct; person to person	Shigellosis
	Rotavirus infection
Food	Salmonellosis
	Campylobacter enteritis
Milk	Brucellosis
	Tuberculosis
Water	Cholera
	Typhoid

Exogenous

These infections are caused by microorganisms acquired from other humans, animals or the environment.

• *Humans:* infections are commonly acquired from other humans who may be infected or carry an organism asymptomatically; carriers may be convalescent (i.e. recovered from the infection, but continuing to excrete the organism), or healthy carriers who have not had an infection (e.g. health-care workers).

• *Animals:* human infections acquired from animals are termed 'zoonoses'. Both domestic and wild animals may be sources of infection; the animal may be infected and carry the microorganism asymptomatically.

• *Environment:* large numbers of microorganisms are found in soil and water, but very few are associated with human infection (e.g. *Clostridium tetani*; *Aspergillus, Legionella*).

Routes of spread of infection

Airborne

Airborne spread is from:

• respiratory droplets from an infected individual (e.g. influenza virus, *Neisseria meningitidis, Mycobacterium tuberculosis*)

• squamous epithelial cells (e.g. *Staphylococcus aureus*)

• fine droplet spread from ventilator nebulizers

(e.g. *Pseudomonas*) or from air-conditioning systems (e.g. *Legionella*)

• spores, e.g. *Aspergillus fumigatus* from contaminated air-handling plants.

Contact transfer (Tables 25.1 and 25.2)

Sources of microorganisms (e.g. respiratory secretion, wound and other skin lesions, urine, faeces) are transferred to an individual either directly or via a vector such as the hands of staff or objects ('fomites'), e.g. surgical instruments, utensils, medical equipment. Other vectors include food, water and insects (e.g. mosquitoes, ticks). Many infections can also be spread via oral–oral or sexual contact. Faecal–oral transmission is a very important mechanism of spread of infection (Table 25.1). Transplacental infection (via placenta from mother to fetus), e.g. rubella, can also occur.

Factors affecting transmission of microorganisms

Survival in the environment

Microorganisms that spread between humans rely on various strategies:

• Respiratory viruses survive poorly outside the host; successful transmission requires production of large numbers of infectious virions. However, the number of virions that are sufficient to cause infection may be small.

• *Mycobacterium tuberculosis* can survive outside the host; transmission can occur with few bacteria.

• *Neisseria gonorrhoeae* dies rapidly outside the host, so transmission is dependent on direct contact.

Establishment on new host

To attach and survive on a new host, a microorganism must combat the physical, chemical and immunological barriers that protect the host at each port of entry (e.g. respiratory tract, gastrointestinal tract, skin, urogenital tract). Microorganisms have a variety of virulence factors that promote attachment and survival (see Chapters 1 and 23).

Strategies to prevent infections

• Reduce exposure to sources of infection:

Table 25.2 Examples of transmission by direct contact or transfer.

TRANSMISSION BY DIRECT CONTACT	
Contact	Infection
Skin	Dermatophyte infection
	Leprosy
Oral	Glandular fever
Sexual	Gonorrhoea
	Chlamydial
	HIV
	HBV
Inoculation injury from infected source	HBV
	HCV
	HIV

• Prevention can be achieved only by knowledge and understanding of the source of the microorganism, the mechanism of transmission and the pathogenesis of infection. These principles can be applied when considering either community-acquired or hospital-acquired (nosocomial) infections and include:
  — reduction in or elimination of sources of infection
  — interruption of transmission
  — prevention of the establishment of infection after exposure to the pathogen by increasing host resistance (e.g. with prophylactic antibiotics, immunization).

Reducing sources of infection

Complete *elimination* of an infection worldwide is difficult and expensive, but has been achieved with smallpox and is aimed for poliomyelitis and measles viruses. Examples of strategies for reducing the source of infection are shown in Table 25.3. The decreased incidence of many infectious diseases in developed countries is associated with improved social conditions and reduced overcrowding; this has resulted in reduced transmission by aerosol, faecal–oral and direct contact transmission.

Preventing transmission of infection

Airborne

Isolation of patients with infections that spread by respiratory droplets is feasible in only a limited

Table 25.3 Examples of reduction in sources of infection.

REDUCTION IN SOURCES OF INFECTION		
Source	Infection	Action
Human	Salmonellosis	Screening of food handlers for *Salmonella* carriage followed by isolation and treatment as appropriate
	Tuberculosis	Isolation; follow-up cases; contact tracing; investigation and treatment of contacts
Animal	Rabies	Quarantine; immunization of pets
	Anthrax	Burning infected dead animals; immunization of animals
Environment	Legionellosis	Improved design and maintenance of humidification systems and air conditioning systems, including chlorination

number of conditions (e.g. tuberculosis (TB)). It is more important in preventing transmission within hospitals.

Faecal–oral

Direct
- Good personal hygiene (handwashing).
- Isolate patients with enteric infections.

Water
- Separate sewage from drinking water.
- Provide potable water (filtration/chlorination) constantly monitored by microbiological testing.

Milk
- Ensure health of cattle.
- Ensure hygienic collection and storage of milk.
- Pasteurize or sterilize milk.
- Monitor microbiology of products.

Food
- Efficient refrigeration of certain foods.
- Separate storage and preparation areas for possibly contaminated foods (e.g. uncooked poultry) from cooked produce.
- Thorough cooking and reheating procedures; complete defrosting before cooking.
- Well-designed catering facilities with high standards of personal hygiene; exclude from work any staff with infections (e.g. staphylococcal skin infections, diarrhoea, salmonella carriers).

- Food contamination is an important public health problem; many cases of food poisoning occur per annum in the UK.

Direct and contact transfer

- Good personal hygiene.
- Sexually transmitted infections can be prevented by: safe sex, contact tracing and health education campaigns.
- Blood-borne virus infections (e.g. hepatitis viruses B and C, and HIV) can be prevented by: sterilization of surgical instruments, safe sharps practice, provision of sterile needles for intravenous drug abusers, and screening of blood and organ donors.

Increasing host resistance

Improved general health

Malnutrition and parasitic infections are important factors that weaken host resistance to other infections.

Antimicrobial prophylaxis

- Surgical prophylaxis is administered mainly in hospitals. Prophylactic antibiotics are important in dental practice to prevent bacterial endocarditis.
- Long-term prophylactic antibiotics are prescribed in a limited number of conditions (e.g. postsplenectomy to prevent pneumococcal infection).
- Short-term prophylactic antibiotics (e.g.

Table 25.4 Examples of sources and microorganisms associated with nosocomial infection.

Type	Common pathogen	Source/spread
Urinary tract infection	Coliforms	Urinary catheterization, patient's own flora
Post operative wound infections	*Staphylococcus aureus*	Patients own flora, environment hand spread, carriage by staff
Ventilation-associated respiratory tract infection	Coliforms	Hospital source or patient's own flora
Intravascular catheter related sepsis	Coagulase-negative Staphylococci	Patient's skin flora

antimalarials) are given when travelling to endemic areas.

• Immunocompromised patients (e.g. transplant recipients and AIDS patients) may receive long-term antimicrobial prophylaxis (e.g. aciclovir to prevent herpes simplex infections; co-trimoxazole to prevent pneumocystis pneumonia).

Immunization

This is dealt with in Chapter 27.

Nosocomial infections

Examples of sources and microorganisms associated with nosocomial infection are given in Table 24.4.

Urinary tract infection

This is the most common nosocomial infection, and is often a complication of catheterization or urinary tract operative procedures. It is a particular problem in urology, obstetric, geriatric, spinal and intensive care units.

The source of the infecting organism may be endogenous or exogenous (cross-infection introduced during urological procedure or manipulation of drainage system). Prevention includes the use of sterile instruments, closed-drainage urinary catheter systems, and careful aseptic technique during the introduction of urinary catheters and subsequent catheter care.

Wound infection

The incidence of wound infection varies according to the type of surgery (Table 25.5). Causative organisms depend on the site and type of operative procedure:

• *S. aureus:* the most common cause after surgery.
• Coliforms and anaerobes: associated most commonly with 'contaminated' or 'infected' operations, mainly involving the gastrointestinal tract.
• Coagulase-negative staphylococci: a frequent cause of device-associated infections (see Chapter 40).
• *Clostridium perfringens:* may result in gas gangrene, now rare in developed countries, but still remaining a problem in developing countries, particularly during wars; gas gangrene may be associated with lower limb amputation for vascular insufficiency.
• β-haemolytic streptococci: less common than staphylococcal infections, but may result in rapid local spread (cellulitis or fasciitis) and septicaemia.

Wound infections are frequently endogenous. Exogenous acquisition may be acquired intraoperatively or postoperatively, either by the airborne route or by direct contact (cross-infection).

Prevention
• Preoperative measures:
— decontamination of skin surfaces immediately preoperatively with appropriate antiseptics
— prophylactic antibiotics for operations where there is a significant risk of wound infection.

Table 25.5 Wound infection relative to surgery undertaken.

WOUND INFECTION		
Type of surgical wound	**Example of surgery**	**Incidence of wound infection (%)**
'Clean': does not involve gastrointestinal, respiratory or genitourinary tracts	Orthopaedic or cardiac surgery	< 5
'Contaminated': involves site with normal flora (apart from skin)	Operations on gallbladder, genitourinary tract	20
'Infected': operation site infected at time of surgery	Emergency surgery for burst, infected appendix	< 50

- Intraoperative measures:
  — modern operating theatres are designed to provide positive pressure ventilation with filtered air so that 'clean' air enters and passes to less clean areas; the ventilation systems reduce the number of microorganisms entering the theatres and, by directing the airflow away from the patient, the number of bacteria-carrying particles generated around any operation wound is minimized
  — the use of sterile instruments, careful hand-washing techniques and the wearing of sterile surgical gloves
  — other procedures to prevent airborne transmission of microorganisms into open wounds include the wearing of headware and close-woven operation gowns
  — good surgical technique to reduce devitalization of tissue and haematoma formation.
- Postoperative measures:
  — 'no-touch' techniques in the cleaning and dressing of wounds have been a major factor in reducing wound infection
  — the use of specially ventilated ward areas for dressing of extensive wounds, such as burns
  — isolation of patients with heavily infected wounds, particularly with antibiotic-resistant organisms.

Lower respiratory tract infection

This is an important cause of morbidity and mortality in hospitalized patients. It is often a complication of intubation (anaesthesia, intensive care) which bypasses the normal physical defences of the respiratory tract (mucus trapping, ciliated epithelium).

Prevention
Many infections are endogenous and isolation of infected patients is of limited value. The use of single-use suction catheters and sterilized anaesthetic circuits with inspiratory and expiratory microbial filters reduces the chance of cross-infection.

Intravascular catheter infections (see Chapter 40)

The use of both peripheral and central vascular catheters (CVCs) is associated with sepsis, ranging from insertion site infections to septicaemia. The source of sepsis is primarily from the patient's own skin flora, with microorganisms gaining access to the catheter either at the time of insertion or via connectors.

Treatment and prevention
See Chapter 40.

Specific microorganisms associated with hospital-acquired infection

Certain organisms, e.g. multi-resistant strains of *Pseudomonas aeruginosa*, meticillin-resistant *S. aureus* (MRSA) and, more recently, vancomycin-resistant enterococci (VRE) need specific control strategies. MRSA is detailed below.

MRSA

See Chapter 41.

Isolation procedures

Hospitals require a policy for isolating patients with certain transmissible infections (source isolation) and for isolating certain immunocompromised patients (e.g. bone marrow transplant recipients) to prevent them acquiring infection (protective isolation). Categories of isolation include:
- *Strict isolation* for highly dangerous conditions, e.g. Lassa fever, viral haemorrhagic fever: these patients require treatment in designated isolation hospitals with special facilities (so-called category IV wards).
- *Source isolation* for conditions such as chickenpox, measles, infections with MRSA and other multiple antibiotic-resistant organisms: includes single-room accommodation with closed door and *negative pressure* to outside; disposable gloves and aprons should be worn when attending the

patient, and hands should be washed before leaving the room. For enteric infections, procedures are required for safe disposal of faeces and urine.

• *Protective isolation*: to protect compromised patients (e.g. bone marrow transplant recipients) from infection. Facilities include single room accommodation, with *positive pressure*, filtered air ventilation. Gowns, aprons and masks need to be worn by all those entering the room; hands should be washed before and after entering.

Infection control teams in hospitals

See Chapter 26.

Chapter 26

Management of infections

Introduction

Infection control is an important aspect of patient care in the community as well as in the hospital environment. Infections that were not present at the time of admission can be both derived in the community (community-acquired infections) or acquired during hospitalization. These hospital-acquired infections are also referred to as nosocomial infections. The source of microorganisms causing infections can be from the patient's own body microflora (endogenous source) or from an external source such as the environment (exogenous source).

Management of the control of infection in hospitals

Each major hospital should have an infection control team and infection control committee. Within the team there are various key members. These are outlined below:

- *Infection control doctor*—a registered medical practitioner with experience in hospital infection; usually a consultant medical microbiologist. Takes a leading role in the effective functioning of the infection control team (see below), and is involved in producing policies and preventive programmes for nosocomial infections.
- *Infection control nurse*—is a registered nurse with additional qualifications in infection control.

Serves as an adviser to the infection control team, providing specialist nursing input into the identification, prevention, monitoring and control of infection.

- *Consultant in communicable disease control (CCDC)*—involved in surveillance, prevention and control of communicable diseases and infections within the boundaries of a health authority. He or she has responsibility for ensuring adequate arrangements for prevention of communicable disease, infection control in the community and will liaise with the hospital infection control team.
- *Senior hospital manager*—ensures that appropriate resources are available and policies and procedures are implemented.

Infection control team

- Responsible for the operational running of an infection control programme.
- The team produces policies and procedures in infection control with regular updates
- Also involved in education of staff in infection control, surveillance of infection
- Provides advice on all matters relating to the prevention and spread of infection.

Infection control committee

Has responsibility for the strategic planning, evaluation for matters relating to infection control. This should involve senior management of the health-care organization. In addition occupational health should be represented on this committee to ensure that the risk of infection to healthcare workers is minimized.

Surveillance

It is important in infection control that a suitable surveillance programme is in place to identify early cases of possible sources of transmissible infections and to prevent spread of infection by ensuring that appropriate infection control measures are implemented. Surveillance can also help in determining trends so that infection control measures can be directed towards particular problems. There are various methods of surveillance including laboratory-based surveillance, e.g. numbers of MRSA (meticillin-resistant *Staphylococcus aureus*) bacteraemias or ward surveillance, e.g. number of postoperative wound infections, numbers of cases of *Clostridium difficile* diarrhoea. Some infections, e.g. MRSA bacteraemias, have to be reported to a National Database.

Major outbreaks

It is important that hospitals as well as the community have in place systems for managing a situation when major outbreaks occur. Medical staff have a statutory duty to notify certain infectious diseases to the CCDC (see Table 26.1), to ensure that any infections that may affect both the hospital and the community are coordinated in terms of control. If an outbreak occurs there are several stages in the investigation, including collection of epidemiological data, screening of hospital cases and putting in place appropriate outbreak control management plan.

Biological warfare

Plans need to be in place both in hospitals and the community in case of contamination with biological material including smallpox and anthrax. Pre-

Table 26.1 List of some of the infectious diseases that need to be notified to a CCDC

Common	Less common
Confirmed measles	Acute poliomyelitis
E. coli O157	Psittacosis
Hepatitis B	Leptospirosis
All meningitis	Relapsing fever
Typhoid & Paratyphoid fever	Yellow fever
Diphtheria	Anthrax
Meningococcal disease	Cholera
Confirmed mumps	Plague
Legionnaires disease	Tetanus
Confirmed rubella	Viral haemorrhagic fever
	Botulism
	Leprosy
	Rabies
	Typhus

cautions include appropriate protective clothing, prophylaxis and immunization.

Prevention of infection to healthcare workers

Healthcare workers can protect themselves from getting infections by several mechanisms:
- Immunisation—healthcare workers should be immunised as appropriate against various infections including hepatitis B.
- Barrier protection—healthcare workers should carry out a risk assessment when dealing with any patient and should assume that a patient may be potentially infectious and apply universal precautions. These are precautions that should be taken for every patient contact to prevent the spread of infection. They include, for example, the use of gloves when directly handling a patient. Avoidance of needlestick injuries is important in the spread of blood-borne viruses.
- Strict isolation—this is required for highly infectious or dangerous pathogens, e.g. Lassa fever and these patients need to be located in designated high-risk isolation hospitals.

Chapter 27

Immunization

The principle of immunization is to increase specific immunity to infection; this may be achieved by the administration of immune serum (passive immunization) or by administration of an antigen that primes the host immune system without causing disease (active immunization).

Passive immunization

- Passive immunization (immunotherapy) gives immediate protection, but immunity lasts for a relatively short time (< 6 months).
- Immunoglobulin for passive immunization is normally obtained from volunteers (e.g. individuals recently immunized) with high titres of specific antibody to a particular organism. However, for infections that are relatively common (e.g. hepatitis A), levels of specific antibody in the general population are high enough to allow the use of pooled plasma from normal blood donors (pooled human immunoglobulin).
- Indications for the use of passive immunization are limited and include:
  — post exposure prophylaxis in immunocompetent hosts when immediate protection is required after exposure to the infection (tetanus, diphtheria, rabies, hepatitis B)
  — post-exposure prophylaxis in immunocompromised hosts (measles, varicella-zoster)
  — therapy (e.g. cytomegalovirus [CMV]

infection in transplant recipients and young children).

Active immunization

- The underlying principle is to stimulate the production of specific B and T lymphocytes (primary response) which develop into memory cells; subsequent exposure to the pathogen then results in the immediate and effective immune response (secondary response) (see Figure 23.4 and Chapter 23).
- Active immunization provides long-lasting protection (often lifelong) against infection, but takes several weeks to become effective.
- Immunization strategies are:
  — to protect susceptible individuals against infection
  — to reduce the incidence of infection in the community (increase 'herd' immunity)
  — to eliminate an infection in a particular country or worldwide (e.g. smallpox).

Vaccines

The ideal vaccine should be:
- able to induce an adequate and appropriate immune response without causing active infection
- safe
- inexpensive
- stable
- easy to administer.

Types of vaccine

Live attenuated

- Live organism with virulence that has been attenuated; this is normally achieved by serial passage in artificial media, e.g. BCG (Bacille Calmette–Guérin) vaccine against tuberculosis; many viral vaccines are live attenuated.
- Advantages: replication of attenuated organism in body closely mimics true infection; generally, only one dose required; inexpensive.
- Complications: random mutations may very rarely occur that result in reversion to virulence; cell lines used for virus production may become contaminated with other viruses (strict quality control procedures are necessary).
- Limitations: live vaccines cannot be given to immunocompromised patients or in pregnancy.
- Successful usage: global eradication of smallpox by use of live vaccinia.

Killed organisms

- Organism killed by chemical or heat treatment, e.g. vaccines against: rabies, influenza, polio (Salk vaccine), *Yersinia pestis*.
- Advantage: can be given to immunocompromised individuals; no reversion to virulence.
- Disadvantage: organism cannot multiply in tissues; several doses required for effective response; expensive.

Toxoids

- Inactivated (formaldehyde treatment) bacterial toxin vaccines, e.g. tetanus, diphtheria.
- Disadvantage: relatively small molecules, so not immunogenic when given alone; require larger molecule (adjuvant) to produce effective immune response.

Subcellular or subviral fractions

- An alternative to using the whole organism is to identify and purify important subunit antigens for use as vaccines.

- Examples include: the polysaccharide capsular vaccines for immunization against pneumococci, meningococci and *Haemophilus influenzae* type b; hepatitis B surface antigen, originally purified from the serum of chronic carriers, but now manufactured by recombinant DNA techniques; and influenza haemagglutinin (HA) and neuraminidase (NA) subunit vaccine.

Individual vaccines

Poliomyelitis

- Both inactivated poliomyelitis vaccine (Salk) and live attenuated vaccine (Sabin) are available.
- Incidence of polio drastically has been reduced in many countries since introduction of routine immunization. By 2004, transmission of wild-type polio virus had been interrupted in the Americas, the western Pacific and European regions.
- In 2004, the UK stopped using the live attenuated vaccine for routine immunization and replaced it with the inactivated vaccine.

Measles/mumps/rubella (MMR)

- Contains live attenuated measles, mumps and rubella viruses.
- Should be given to children irrespective of history of measles, mumps or rubella.

Influenza

- Vaccine regularly updated to reflect current strains in circulation. Contains two type A and one type B virus strains.
- Gives 70% protection; protects for approximately 1 year.
- Recommended for patients > 65 years of age, people at any age with chronic respiratory disease, heart disease, metabolic diseases, liver disease or immunosuppression, and for people living in long-stay care facilities.

Hepatitis A

- Formaldehyde-inactivated vaccine made from a strain of hepatitis A virus.

- Recommended for travel to endemic areas, for control of outbreaks, patients with chronic liver disease or haemophilia, homosexual individuals and laboratory staff who work with the virus.

Hepatitis B

- Both active and passive immunization available:
  — vaccine: contains hepatitis B surface antigen (HBsAg) prepared by recombinant DNA techniques; recommended for at-risk groups, e.g. healthcare workers, neonates born to HBV carrier mothers, close family contacts of cases or a carrier and patients with chronic renal failure
  — specific immunoglobulin for passive immunisation; recommended for post-exposure prophylaxis for non-immune individuals and neonates born to hepatitis B 'e' antigen (HBeAg)-positive carrier mothers.

Human varicella-zoster virus

- Both active and passive immunization available:
  — varicella-zoster immunoglobulin (ZIG): prepared from pooled plasma of blood donors with recent chickenpox or herpes zoster; recommended use for post-exposure prophylaxis in non-immune pregnant immunocompromised patients, infants < 1 weeks of age without maternal antibody protection; non-immune pregnant women newborns when mother develops chickenpox (1 week before or after delivery)
  — vaccine: live attenuated vaccine; in some countries (e.g. the USA) it is part of the routine childhood immunization programme; in the UK it is recommended only for non-immune healthcare workers.

Rabies

Active immunization (killed vaccine) is used for pre-exposure prophylaxis, e.g. in animal handlers, laboratory staff working with the virus and long-term travellers to countries where rabies is endemic and access to medical care may be difficult. Both passive (rabies immunoglobulin) and active (killed vaccine) immunization are available for post-exposure prophylaxis.

Yellow fever

Live attenuated vaccine for travellers to endemic areas and laboratory workers handling infected material.

Tuberculosis

- BCG vaccine contains a live attenuated strain derived from *Mycobacterium bovis*. This stimulates hypersensitivity to *M. tuberculosis*.
- Should not be given to immunosuppressed patients.
- Routinely given in childhood in some countries.
- Must be given intradermally by skilled operator. The reaction at the injection site must be assessed to ensure efficacy.

Streptococcus pneumoniae

- Polyvalent vaccines containing purified capsular polysaccharide from either 23 capsular types of pneumococci (accounts for 90% of isolates in the UK) or 7 capsular types conjugated to protein.
- Recommended for patients with asplenia or dysfunctional spleens.
- 23-valent vaccine used in at-risk patients aged > 5 years and the 7-valent conjugate vaccine in children aged < 5 years.

Cholera

Oral vaccine contains a mixture of two heat-killed serotypes and the recombinant B subunit of cholera toxin. Protection lasts 3–6 months; recommended for travel to endemic areas. Should be given at least 1 week before travel.

Typhoid

- Vaccine is the Vi capsular polysaccharide from *Salmonella typhi*.
- Recommended for travel to endemic areas.

Meningococcal infection

- A purified, heat-stable, lyophilized extract from the polysaccharide outer capsule of *Neisseria meningitidis*.
- Effective against serogroups A and C; no vaccine available yet for group B organisms.
- Recommended for close contact of cases of meningococcal infection.

Diphtheria

- Toxin of *Corynebacterium diphtheriae*; rendered non-toxigenic (toxoid), but retains antigenicity.
- Disease virtually eliminated in the UK and other developed countries.
- Booster doses use 'low-dose' vaccine.

Pertussis

- Acellular pertussis vaccine derived from highly purified components of *Bordetella pertussis*.
- Acellular vaccines are much less likely to be associated with severe neurological illness in children than a whole cell vaccine.

Tetanus

- Prepared by treating a cell-free preparation of toxin with formaldehyde (tetanus toxoid).
- Part of childhood immunization schedules in many countries.
- Booster doses often given when patients present with contaminated wounds.
- Passive immunization with immunoglobulin is sometimes used in severely contaminated wounds.

Haemophilus influenzae type B (Hib)

- Capsular polysaccharide linked to protein.
- *H. influenzae* type b infection occurs principally in children aged < 4 years; immunization is normally given to infants during first 6 months of life.
- Also given to patients with asplenia or splenic dysfunction.

Japanese encephalitis

A killed vaccine available for travellers to endemic areas and for laboratory workers at risk of exposure to the virus.

Administration of vaccine

- Childhood immunization schedules are common in most developed countries and also in developing countries (World Health Organization [WHO] extended programme of immunization—EPI). Timings reflect the need to protect children against important infections early in childhood,

Table 27.1 Childhood immunization schedule in the UK (2005).

CHILDHOOD IMMUNIZATION	
Age	**Vaccine**
2–4 months (three doses at 4-week intervals)	Diphtheria, tetanus, acellular pertussis (DTP)
	Inactivated polio
	Haemophilus influenzae type b (Hib)
	Meningococcal C
12–18 months	Measles, mumps, rubella (MMR)
4–5 years (school entry)	Diphtheria, tetanus, acellular pertussis booster
	Inactivated polio booster
	MMR booster
13–18 years (school leaving)	Inactivated polio
	Diphtheria (low dose), tetanus

the need to give booster doses for some vaccines (particularly toxoids), and the need to immunize at convenient times when take-up rates will be high. The UK immunization schedule (2005) is shown in Table 27.1.

- Adults require immunization:
  - to boost childhood immunization (e.g. tetanus)
  - to cover travel to endemic areas (e.g. cholera, typhoid, yellow fever, polio, meningitis, hepatitis A)
  - for protection in high-risk occupations (e.g. hepatitis B in healthcare workers; rabies in laboratory workers)
  - for protection of at-risk groups (pneumococcal vaccine for post-splenectomy patients; influenza vaccine for elderly people and patients with chronic respiratory/cardiac conditions).

- To maintain their efficacy, all vaccines and immunological products need to be carefully stored under the conditions described in product literature.

Chapter 28

Sterilization and disinfection

Definitions

- *Sterilization*: the process whereby all viable microorganisms, including spores, are removed or killed.
- *Disinfection*: the process whereby most, but not all, viable microorganisms (excluding spores) are removed or killed.
- *Pasteurization*: the process used to eliminate pathogens in foods such as milk. Spores are unaffected.

Sterilization and disinfection

Methods of sterilization and disinfection employed in clinical practice are related to the material being treated. Sterilization and disinfection can be divided into either physical or chemical methods. The efficacy of these methods is influenced by the processing time, the presence of organic materials (e.g. blood) and the material being treated. It is essential, before any sterilization or disinfection process, that items are thoroughly cleaned to remove organic matter

Methods

Dry heat

Recommended for sterilization of heat-tolerant articles; dry heat kills microorganisms by oxidation; rapid, reliable; best method for heat-resistant articles.

- *Flaming to red heat*: metal instruments, such as dental reflection mirrors, can be sterilized by direct heating. In an emergency, scalpels can be sterilized by dipping the blade in methylated spirit and burning off the spirit.
- *Hot air oven*: usually operates at 160°C for 1 h to sterilize items such as glassware; time must be allowed for all items in the oven to reach 160°C before the 1 h exposure is timed.
- *Incineration*: complete burning of material in an incinerator; this is used for the safe disposal of items such as contaminated dressings, pathology specimens and laboratory cultures.

Moist heat

Sterilization by moist heat is more rapid and efficient than that by dry heat; the presence of water causes protein denaturation, resulting in disruption of cell membranes, and improves heat penetration.

- *Autoclaving*: autoclaves work on the same principle as the domestic pressure cooker. Saturated steam is produced under pressure, resulting in temperatures > 100°C; this ensures total destruction of all microorganisms, including spores. Contaminated items are exposed initially to steam under pressure and, when the required temperature is reached, steam is evacuated to produce a vacuum and dry the load. A typical cycle would be 15 pounds/square inch (101 kPa) at 121°C for 15 min. Autoclaving is the most common method of sterilization in hospitals. To ensure efficient sterilization, the process must be carefully controlled:

— each sterilization cycle must be monitored by continuous recording of the temperature and steam pressure

— items must be loaded into the autoclave to allow even and complete exposure to the steam; they should be covered to prevent recontamination after removal from the autoclave

— efficacy of the sterilization can be tested by several methods: Browne's sterilizer control tubes contain a chemical that changes colour when exposed to various temperatures; spore strips (*Bacillus stearothermophilus* spores are normally used) can be placed among the items being autoclaved; survival of the spores after autoclaving indicates a problem with the process; tapes that change colour after adequate treatment are placed on each article; thermocouples are placed in typical loads.

• *Boiling water*: boiling water baths are still used in the clinical situation (e.g. cleaning of instruments). However, they cannot be relied upon to sterilize instruments, because spores will survive boiling.

• *Pasteurization*: this is used to destroy vegetative bacteria (e.g. *Mycobacterium*, *Salmonella*, *Brucella* species) in milk by heating to 63°C for 30 min (holder method) or 72°C for 20 s (flash method). Spores survive pasteurization.

Filtration

Filtration can be used to remove microorganisms from fluids that cannot tolerate heat. However, filtration is slow and many viral particles can pass through filters. Filtration of air is important in operating theatres and in pharmacy departments where sterile solutions are being prepared. These filters remove most microorganisms, but do not result in sterile air. Current guidance recommends a certain quality of air in some of these areas.

Radiation

Many bacteria are killed rapidly on exposure to radiation. Gamma radiation, usually from a cobalt-60 source, is being used extensively in the commercial sector for the sterilization of materials such as plastic disposable syringes. This process does not result in a rise in temperature of the materials, but the time required for sterilization can be as long as 48 h. The equipment and safety procedures required for sterilization by gamma radiation limit its use within hospitals.

Chemicals

A variety of chemicals are used to kill bacteria and their spores; these are classified into disinfectants or antiseptics:

• *Disinfectant*: a chemical used to remove or kill most viable organisms present (Table 28.1).

• *Antiseptic*: disinfectant used for skin cleaning; does not sterilize skin; less irritant compared with disinfectant.

Some disinfectants, e.g. glutaraldehyde, have been used to sterilize equipment such as fibreoptic endoscopes. Glutaraldehyde is being replaced, for this purpose, with peracetic acid or chlorine dioxide, because of the dangers to users.

Gases, such as ethylene oxide, are used to sterilize single-use medical items; however, they may be toxic and explosive, as ethylene oxide is. Formaldehyde gas is used to disinfect microbiology containment laboratories after contamination episodes.

The choice of disinfectant for different hospital applications is made on the basis of the disinfectant's antimicrobial activity, inactivation by organic material, toxicity and compatibility with the surface or equipment on which it is to be used. Most hospitals formulate a disinfection policy that specifies the agent, the appropriate concentration and the correct contact time for different situations.

Prions

The stable structure of prions makes it very difficult

Table 28.1 Properties of commonly used disinfectants.

PROPERTIES OF DISINFECTANTS

Group	Example (notes)	Bactericidal Gram negative	Bactericidal Gram positive	Sporicidal	Fungicidal	Viricidal	Mycobactericidal	Inactivated by organic matter	Human tissue toxicity	Uses
Alcohols	70% ethyl alcohol	+	+	–	–	–	–	±	±	Skin antiseptic, 100% alcohol ineffective; need addition of water to 70% v/v; rapid action
Aldehydes	Formaldehyde gas or aqueous forms	+	+	+	+	+	+	±	+	Fumigation
	Glutaraldehyde	+	+	+	+	+	+	±	+	Disinfection of fibreoptic endoscopes; staff exposure to fumes
Biguanides	Chlorhexidine	–	+	–	–	–	–	+	–	Hand wash; skin antiseptic
Halogens	Hypochlorites	+	+	±	+	+	–	+	±	General environmental cleaning; blood spills; treating water
	Chlorine									Decreased activity on storage
	Chlorine dioxide	+	+	+	+	+	+	+	+	
	Iodine	+	+	±	+	+	–	+	±	With alcohol, used for skin preparation; hand wash and skin ulcers

								Uses
Peracetic acid	+	+	+	+	+	+	+	Disinfection of fibreoptic endoscopes
Phenolics	Phenol (carbolic acid)	±	+	−	−	+	+	+	Absorbed by rubber; too irritant for general use
	Clear phenolics	±	+	·	−	+	−	+	Decontamination of floors, etc.
	Hexachlorophane	−	+	−	−	−	−	−	Powder form for skin application, skin disinfection
	Chloroxylenols (Dettol)	−	±	−	−	−	+	±	
Quaternary ammonium compounds	Cetrimide (bacteriostatic)	±	+	−	±	−	+	−	Skin disinfection
	Benzalkonium chloride	±	+	−	+	−	+	−	Preservative of topical preparation/ antimicrobial; plastic catheters

+, yes; −, no; ±, intermediate.

151

for their infectivity to be destroyed by the conventional means described above. Autoclaving at 134°C for 1 hour does not prevent infectivity of all forms of prions. High concentrations of hypochlorite for 60 min are effective but corrode most instruments. The use of aldehydes increases prion infectivity. If surgical instruments are in contact with tissues that are likely to contain prions, they should be disposed of after use.

Chapter 29

Antibacterial agents

Definitions

Antimicrobial agent

Substance with inhibitory properties against microorganisms (includes antibiotics and synthetic compounds) but with minimal effects on mammalian cells ('selective poisons').

Antibiotic

Substance produced by microorganisms that inhibits the growth of (bacteriostatic) or kills (bactericidal, viricidal, fungicidal) other microorganisms. The term 'antibiotic' is often incorrectly used to include all antimicrobial agents, some of which are synthetic, e.g. sulphonamides.

Semi-synthetic antibiotics

Antibiotics chemically altered to improve properties, e.g. stability or spectrum of activity.

Mechanism of action

Antimicrobial agents are divided into groups according to their mode of action. The potential targets for antimicrobial action include the bacterial cell wall, cell membrane, protein synthesis and nucleic acid synthesis (Table 29.1).

Inhibitors of cell-wall synthesis

Most bacteria, unlike mammalian cells, have a cell wall. The main groups of antimicrobial agents that act selectively on the bacteria cell wall are the β-lactams and glycopeptides.

β-Lactam antibiotics

Structure

This is a large group of compounds, all with a β-lactam ring (Figure 29.1). The structure of the side chains and rings attached to the β-lactam ring determines the class of antibiotic and also its properties. Cephalosporins have a six-membered ring, whereas penicillins have a five-membered ring. Many modifications have been made to the penicillins and cephalosporins by addition of structural groups at various sites.

Action

These antibiotics interfere with the cross-linking of the cell-wall polymer peptidoglycan by inhibiting carboxypeptidase and transpeptidase (penicillin-binding proteins [PBPs]) reactions, which form a link between the building blocks of peptidoglycan, *N*-acetylmuramic acid and *N*-acetylglucosamine; this weakens the cell wall and results in rupture (lysis) of the microorganism (Plate 53). The antibiotics within this group include the penicillins (e.g. benzylpenicillin), cephalosporins (e.g. cefuroxime) and carbapenems (e.g. meropenem).

Penicillins

The penicillins comprise a large group of mainly semi-synthetic compounds based on benzylpenicillin. They can be divided into the following groups:

1 Benzylpenicillin, active against many streptococci.
2 Orally absorbed penicillins, similar to benzylpenicillin (e.g. penicillin V).
3 Penicillins resistant to staphylococcal β-lactamase (e.g. flucloxacillin).
4 Extended spectrum penicillins, with activity against streptococci and many coliforms (e.g. ticarcillin).

Table 29.1 Sites of action of different antimicrobial agents.

SITES OF ACTION	
Site of action	**Examples of antimicrobials**
Cell wall	Penicillins
	Cephalosporins
	Vancomycin
Cell membrane	Polymyxins
Protein synthesis	Aminoglycosides
	Chloramphenicol
	Fusidic acid
	Macrolides
	Tetracyclines
DNA synthesis	Sulphonamide
	Trimethoprim
	Quinolones
RNA synthesis	Rifampicin

5 penicillins with activity against *Pseudomonas aeruginosa* (e.g. piperacillin).

Pharmacokinetics
The penicillins have a wide distribution in the body. Excretion is mainly renal.

Toxicity
Hypersensitivity reactions include anaphylaxis and urticarial skin rashes (maculopapular); 10% of penicillin-allergic patients are also allergic to cephalosporins.

Antibacterial resistance
• Alteration in target site, e.g. meticillin-resistant *Staphylococcus aureus* (MRSA) can produce a PBP with low affinity for β-lactams and therefore continue to function in the presence of the antibiotic.
• β-Lactamase production: this bacterial enzyme hydrolyses the β-lactam ring; some β-lactams are unaffected by β-lactamases (i.e. are β-lactamase stable).
• Cell membrane alterations, reducing uptake or increasing loss from the cell by drug efflux pumps in Gram-negative bacteria.

Examples of penicillins
Benzylpenicillin
• *Pharmacokinetics*. Unstable in acid and destroyed in the stomach; can be given intravenously (i.v.) or intramuscularly (i.m.). Widely distributed in the body. Excretion is predominantly renal.
• *Antimicrobial spectrum of activity*. Effective mainly on Gram-positive organisms and Gram-

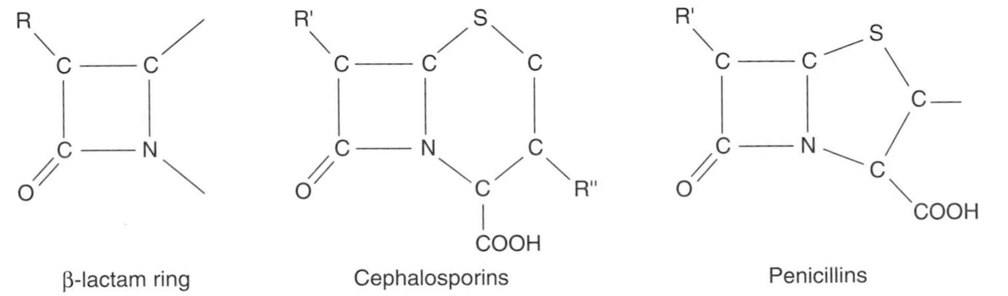

Figure 29.1 Structure of β-lactam antibiotics.

negative cocci, including streptococci, neisseriae and *Treponema pallidum*.

• *Clinical applications*. Streptococcal infections, pneumococcal pneumonia, meningococcal meningitis, gonorrhoea and syphilis.

Flucloxacillin

• *Pharmacokinetics*: this is a semi-synthetic penicillin. It is well absorbed orally; part metabolized in the liver, part excreted in urine.

• *Antimicrobial spectrum of activity*.: flucloxacillin is active against both β-lactamase-positive and -negative strains of *S. aureus*, *Streptococcus pyogenes*.

• *Clinical applications*: *S. aureus* infections, including abscesses, pneumonia, endocarditis and septicaemia.

Ampicillin and amoxicillin

• *Pharmacokinetics*: semi-synthetic penicillins; acid stable and well absorbed orally. Excretion is predominantly renal. Amoxicillin is derived from a minor chemical change in ampicillin, which improves oral absorption.

• *Antimicrobial spectrum of activity*: similar to benzylpenicillin, but more active against *Enterococcus faecalis*, *Haemophilus influenzae* and some Gram-negative aerobic bacilli; activity against many β-lactamase-producing bacteria can be enhanced by the co-administration of β-lactamase inhibitors, such as clavulanic acid.

• *Clinical applications*: urinary and respiratory tract infections caused by *H. influenzae* and some coliforms; *Listeria monocytogenes* meningitis; infections with enterococci.

Piperacillin

• *Pharmacokinetics*: a semi-synthetic penicillin, piperacillin is not absorbed orally. The principal route of excretion is via the kidneys.

• *Antimicrobial spectrum of activity*: piperacillin is moderately active against many Gram-positive organisms, neisseriae and *H. influenzae*. It has wider activity than ampicillin against coliforms and is active against *Pseudomonas aeruginosa*.

• *Clinical application*: severe Gram-negative aerobic bacillary infections (e.g. septicaemia, peritoni-

tis, neutropenic sepsis) caused by coliforms and *P. aeruginosa*.

Cephalosporins

The cephalosporins are a large group of antimicrobial agents based on cephalosporin C with a structure similar to the penicillins (see Figure 29.1). They are classified as:

1 First generation: early compounds.

2 Second generation: compounds resistant to β-lactamases.

3 Third generation: compounds both resistant to β-lactamases and with an increased spectrum of activity.

Antimicrobial spectrum of activity

Variable but generally include *S. aureus* (including β-lactamase-positive strains), streptococci, many Gram-negative species, including neisseriae, *Haemophilus* and coliforms.

Toxicity/side effects

Hypersensitivity with rashes, usually maculopapular. Cross-reactivity (about 10%) with penicillin allergy.

Resistance

Similar to the penicillins.

Examples of cephalosporins

Cefuroxime

• *Pharmacokinetics*: available in oral and parenteral forms. Wide distribution in the body with renal excretion.

• *Antimicrobial spectrum of activity*: *S. aureus*, most streptococci, coliforms. (Note that *P. aeruginosa*, enterococci and anaerobes are resistant.)

• *Clinical applications*: urinary tract infections, soft tissue and chest infections, septicaemia.

Ceftazidime

• *Pharmacokinetics*: not absorbed orally. Given i.m. or i.v. and is distributed into many body tissues and fluids. Excretion is exclusively renal.

• *Antimicrobial spectrum of activity*: less activity against Gram-positive bacteria but wide activity against Gram-negative aerobic bacteria including *P. aeruginosa*.

- *Clinical application*: severe Gram-negative aerobic bacillary infections (e.g. septicaemia, peritonitis) caused by coliforms and *P. aeruginosa*.

Carbapenems

The carbapenems have a structure similar to penicillins.

Antimicrobial spectrum of activity
Broad activity against Enterobacteriaceae (including those producing extended-spectrum β-lactamase), Gram-positive bacteria and anaerobes.

Toxicity/side effects
Hypersensitivity with rashes, usually maculopapular. Cross-reactivity (about 10%) with penicillin allergy.

Resistance
As a result of hydrolysis by carbapenemases; reduced uptake by cell (this may be acquired during treatment).

Examples of carbapenems
Ertapenem
- *Pharmacokinetics*: must be given i.v.; once daily dosing. Renal excretion.
- *Antimicrobial spectrum of activity*: broad activity (see above) but poor activity against non-fermenting Gram-negative bacilli, e.g. *Pseudomonas* sp. and *Acinetobacter*.
- *Clinical application*: severe Gram-negative bacillary infections, (e.g. septicaemia, peritonitis) if *Pseudomonas* or *Acinetobacter* spp. are not likely to be present. Once daily dosing may be useful in treating patients in the community for urinary tract infections caused by multi-resistant Enterobacteriaceae.

Imipenem with cilastatin, and meropenem
- *Pharmacokinetics*: must be given i.v. Renal excretion. Cilastatin is combined with imipenem to inhibit renal dehydropeptidase I and improve imipenem blood levels. Imipenem/cilastatin is

more likely to provoke convulsions than meropenem. Meropenem is thus favoured for treating CNS infections.
- *Antimicrobial spectrum of activity*: includes the 'non-fermenters' *Pseudomonas spp.* and *Acinetobacter spp.*
- *Clinical application*: severe Gram-negative bacillary infections, including septicaemia, peritonitis, empirical treatment of neutropenic sepsis.

Glycopeptides

The glycopeptide antibiotics, vancomycin and teicoplanin, interfere with peptidoglycan assembly, but at a different site from β-lactams, resulting in cell lysis and death.

Pharmacokinetics
Not absorbed from the gastrointestinal tract. Glycopeptides are given i.v. and are widely distributed; largely excreted by the renal route; poor penetration into cerebrospinal fluid (CSF).

Mechanism of action
Interfere with cell-wall synthesis by binding to the pentapeptide chain, preventing incorporation of new subunits into the cell wall.

Antimicrobial spectrum of activity
Glycopeptides are bactericidal against Gram-positive bacteria, including staphylococci. They have no activity against Gram-negative bacteria.

Toxicity
Vancomycin requires monitoring of serum levels because of ototoxicity and nephrotoxicity.

Resistance
Occasionally seen in enterococci. *S. aureus* of reduced sensitivity or of true resistance to vancomycin has been described rarely.

Clinical applications
Severe staphylococcal infections (especially MRSA), including endocarditis, peritonitis; infections of prosthetic devices caused by coagulase-negative staphylococci; oral treatment of

Clostridium difficile-associated pseudomembranous colitis.

Newer antibiotics for use against resistant Gram-positive bacteria

The increasing importance of resistance in Gram-positive bacteria (particularly *S. aureus* and enterococci) has led to the development of some new antibiotics. These include linezolid (an oxazolidinone), daptomycin (a lipopeptide) and Synercid® (a mixture of the streptogramins, quinupristin and dalfopristin).

Linezolid®
- *Pharmacokinetics*: excellent oral absorption; oral dose is equivalent to intravenous. Can be used in renal failure. Good penetration of skin and soft tissues, bones and joints, and the central nervous system (CNS).
- *Mechanism of action*: inhibits bacterial protein synthesis.
- *Antimicrobial spectrum of activity*: activity against a wide range of Gram-positive bacteria including MRSA and glycopeptide-resistant enterococci.
- *Toxicity*: following prolonged use (> 2 weeks) may cause thrombocytopenia and anaemia (check blood counts). Interaction with monoamine oxidase inhibitors.
- *Resistance*: found rarely in *S. aureus* and enterococci. Usually the result of ribosomal mutations. May be acquired during treatment.
- *Clinical applications*: treatment of infections caused by MRSA and glycopeptide-resistant enterococci, including skin and soft tissue infections; ventilator-associated MRSA pneumonia; bone and joint, and CNS infections.

Antimicrobial agents acting on cell membranes

Only a few agents target the microbial membrane. These include polymyxins and the polyene and imidazole antifungals. Polymyxins (e.g. colistin) are similar to detergents and disrupt the membrane, resulting in loss of cytoplasmic content. They are nephrotoxic and usually used as topical agents, e.g. in gut sterilization regimens. Colistin is very occasionally used to treat multi-resistant Gram-negative bacilli, e.g. *Acinetobacter* spp.

Inhibitors of protein synthesis

This group includes the aminoglycosides, macrolides, chloramphenicol and tetracycline, which target the difference between human and bacterial ribosomes, the latter having 30 and 50 S subunits:
- Aminoglycosides (e.g. gentamicin and tobramycin) inhibit protein synthesis by blocking the formation of the initiation complex after mRNA has been utilized.
- Macrolides (e.g. erythromycin) bind to the 50 S subunit of ribosomal RNA and inhibit the formation of initiation complexes.
- Chloramphenicol binds to the 50 S subunit and interferes with the linkage of amino acids in the peptide chain formation.
- Tetracyclines bind to the 30 S subunit, preventing binding of aminoacyl-transfer RNA to the acceptor site in the ribosome, thereby inhibiting amino acid chain elongation.

Aminoglycosides

Examples
Gentamicin, tobramycin, netilmicin, amikacin, streptomycin.

Pharmacokinetics
Poorly absorbed from the gut; they have poor penetration into tissue and fluids; excretion is almost entirely by the kidneys. Serum levels require monitoring with careful dosage adjustment particularly in renal failure.

Antimicrobial spectrum of activity
Active against staphylococci and Gram-negative aerobic bacilli including *Pseudomonas* spp. they have no activity against streptococci and anaerobes. They are bactericidal, showing synergy with β-lactams.
- *Toxicity/side effects.* Hypersensitivity, ototoxicity and nephrotoxicity.
- *Resistance.* Mechanisms include changes in ribo-

somal binding of the drug, decreased permeability, increased efflux and, most importantly, inactivation by aminoglycoside-modifying enzymes.

● *Clinical applications.* Severe sepsis caused by coliforms and other Gram-negative aerobic bacilli.

Macrolides and lincosamides

Macrolides
Examples
Erythromycin, clarithromycin and azithromycin.

Pharmacokinetics
Absorbed following oral administration; also given i.v. They are well distributed and penetrate into phagocytic cells. Excretion is mainly in the bile.

Antimicrobial spectrum of activity
Most Gram-positive organisms, *Haemophilus*, *Bordetella* and *Neisseria* spp. and some anaerobes. Many coliforms are resistant. Active against chlamydiae, rickettsiae and mycoplasmas.

Toxicity/side effects
Gastrointestinal upsets, rashes, hepatic damage (rare).

Resistance
This is by alteration of the RNA target which results in cross-resistance with clindamycin, or drug efflux (with no cross-resistance to clindamycin).

Clinical applications
Streptococcal and staphylococcal soft tissue and bone infections; respiratory infections caused by *Haemophilus influenzae*, mycoplasmas, legionellae, chlamydiae and campylobacter enteritis.

Lincosamides
Examples
Lincomycin and clindamycin.

Pharmacokinetics
Usually given orally, but can be given i.v. or i.m. Penetrates well into the bone; metabolized in the liver.

Antimicrobial spectrum of activity
Similar to erythromycin, but more active against anaerobes.

Toxicity
Associated with pseudomembranous colitis, but probably no more than many other antibiotics.

Resistance
See erythromycin.

Clinical applications
They are used in staphylococcal infections, particularly osteomyelitis and cellulitis, as an alternative to flucloxacillin, and in anaerobic infections.

Chloramphenicol

Pharmacokinetics
Rapidly absorbed after oral administration with good penetration into many tissues including the CSF and brain. Metabolized in the liver to inactive metabolites, which are excreted via the kidney.

Antimicrobial spectrum of activity
Effective against a wide range of organisms, including Gram-negative and Gram-positive bacteria, chlamydiae, mycoplasmae and rickettsiae.

Toxicity/side effects
Chloramphenicol has a dose-related, but reversible, depressant effect on bone marrow; rarely, irreversible, potentially fatal, marrow aplasia. Toxicity in neonates causes grey baby syndrome.

Resistance
This occurs as a result of inactivation by an inducible acetylase enzyme; reduced permeability.

Clinical applications
Meningitis, typhoid, brain abscess; topical for eye infections.

Tetracyclines

Pharmacokinetics
Can be given orally or i.v. Penetrate well into body

fluids and tissues as a result of lipid solubility. Excretion is via the kidney and bile duct. Doxycycline may be given once a day.

Antimicrobial spectrum of activity
Broad spectrum of activity against many Gram-positive and some Gram-negative bacteria, chlamydiae, mycoplasmae and rickettsiae. All tetracyclines are essentially equivalent in spectrum of activity. Tigecycline is a new tetracycline that may be useful against multi-resistant, non-fermenting, Gram-negative bacilli, such as *Acinetobacter* spp.

Toxicity/side effects
Gastrointestinal intolerance; deposition in developing bones and teeth precludes use in young children and in pregnancy; hypersensitivity occasionally, with skin rashes.

Resistance
Efflux from cell; less commonly by protection of target from the tetracycline.

Clinical application
Important in treatment of infections caused by mycoplasmas, rickettsiae and chlamydiae.

Inhibition of nucleic acid synthesis

Several classes of antimicrobial agents act on microbial nucleic acid, including the quinolones, sulphonamides, diaminopyrimidines, rifampicin, and nitroimidazoles.

Quinolones

Examples
Nalidixic acid; fluoroquinolones with activity against coliforms and other Gram-negative bacilli, e.g. ciprofloxacin, ofloxacin and norfloxacin; some fluoroquinolones also have increased activity against pneumococci, e.g. levofloxacin and moxifloxacin.

Mechanism of action
Inhibit the action of bacterial DNA gyrases

(topoisomerases) which are important in 'super-coiling' (folding and unfolding DNA during synthesis).

Nalidixic acid

Pharmacokinetics
Rapidly absorbed after oral administration and almost entirely excreted in the urine.

Antimicrobial spectrum of activity
Bacteriostatic against a wide range of Enterobacteriaceae; no activity against *Pseudomonas aeruginosa* or Gram-positive organisms.

Toxicity
Gastrointestinal and neurological disturbances.

Resistance
Changes in DNA gyrases result in reduced affinity for nalidixic acid; efflux from the cell; less commonly may be caused by target protection. May be acquired during treatment as a result of chromosomal mutation.

Clinical application
Simple urinary tract infections.

Fluoroquinolones

Pharmacokinetics
Generally good absorption after oral administration; penetrate well into body tissues and fluids. Eliminated by renal excretion and liver metabolism, with some excretion in bile.

Antimicrobial activity
Fluoroquinolones with a Gram-negative spectrum, e.g. ciprofloxacin, have a wider spectrum of activity than nalidixic acid; includes *Pseudomonas* spp, Enterobacteriaceae, legionellae, chlamydiae, mycoplasmas and rickettsiae; poor activity against pneumococci. Fluoroquinolones with better Gram-positive activity (moxifloxacin, levofloxacin) have activity against pneumococci. None of the fluroquinolones has any useful activity against streptococci, enterococci or anaerobes.

Toxicity and side effects

Gastrointestinal disturbances, photosensitivity, rashes, neurological disturbances, including seizures (rare) and ruptured Achilles' tendon. Possible effect on growing cartilage relatively contraindicates use of fluoroquinolones in children.

Clinical applications

For fluoroquinolones with Gram-negative spectrum: urinary tract infections, gonorrhoea, respiratory and other infections caused by Gram-negative aerobic bacilli, including *Pseudomonas aeruginosa*; also effective for enteritis caused by shigellae and salmonellae; for prophylaxis in contacts of meningococcal disease. For fluoroquinolones with more Gram-positive spectrum: community-acquired pneumonia, exacerbation of chronic obstructive pulmonary disorder (COPD), sinusitis, otitis media.

Sulphonamides

Examples

Sulphadiazine, sulphadimidine.

Mechanism of action

Act on folic acid synthesis as competitive antagonists of *p*-aminobenzoic acid, inhibiting purine and thymidine synthesis (Figure 29.2). Trimethoprim also acts on the folic acid synthesis in the next step in the metabolic pathway.

Pharmacokinetics

Well absorbed after oral administration. Distributed throughout the body and excreted mainly in the urine.

Antimicrobial spectrum of activity

Broad spectrum of activity, including streptococci, neisseriae, *H. influenzae*.

Toxicity.

Hypersensitivity reactions, causing renal damage, rash; rarely Stevens–Johnson syndrome (a serious form of erythema multiforme); bone marrow failure.

Resistance

Many Enterobacteriaceae are now resistant, as a result of production of an altered dihydropteroate synthetase with decreased affinity for sulphonamide.

Clinical applications

Limited by resistance and toxicity but include nocardiasis and in combination with trimethoprim (co-trimoxazole) for the prevention and treatment of *Pneumocystis jiroveci* (previously named *P. carinii*) pneumonia.

Diaminopyrimidines

Trimethoprim

Mechanisms of action

Prevents synthesis of tetrahydrofolic acid by inhibiting dihydrofolate reductase (Figure 29.2).

Pharmacokinetics

Rapidly absorbed after oral administration. Excretion is via urine.

Antimicrobial spectrum of activity

Broad spectrum of activity against many Gram-positive and Gram-negative bacteria. *Pseudomonas* spp. are resistant.

Toxicity

Folate deficiency.

Resistance

Mainly as a result of modification or production of multiple copies of the target enzyme, dihydrofolate reductase.

Clinical applications

Urinary tract and respiratory infections.

Co-trimoxazole (trimethoprim combined with sulphamethoxazole)

This was prescribed commonly in general practice for the treatment of urinary tract and respiratory infections; however it is now recognized that trimethoprim alone is probably just as effective

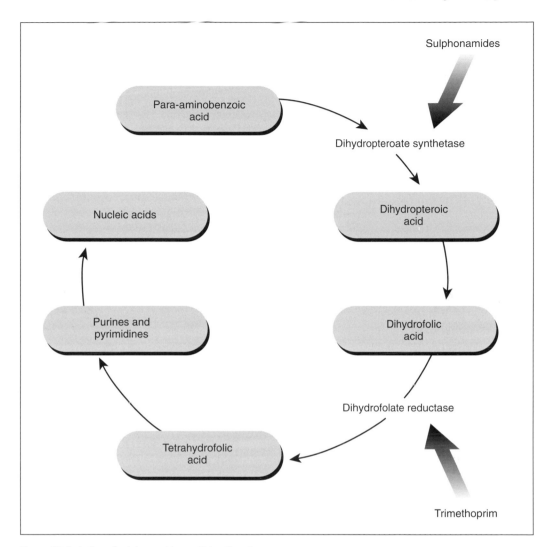

Figure 29.2 Action of sulphonamides and trimethoprim.

and does not have the toxicity problems associated with sulphonamides.

Rifamycins

Rifampicin
Mechanism of action
Binds to RNA polymerase and blocks synthesis of mRNA.

Pharmacokinetics
Well absorbed after oral administration; widely distributed in the body, including CSF. Metabolized in the liver and elimination primarily by biliary secretion.

Antimicrobial spectrum of activity
Active against many staphylococci, streptococci, *H. influenzae*, neisseriae, legionellae and mycobac-

teria. Enterobacteriaceae and *P. aeruginosa* are resistant.

Toxicity

Adverse reactions include skin rashes and transient liver function abnormalities; a rare cause of hepatic failure. Potent inducer of hepatic enzymes, reducing the effect of warfarin and oral contraceptives.

Resistance

Resistant mutants emerge when used as a single agent; resistance is the result of the change in a single amino acid of the RNA polymerase target site. Often used in combination with other agents to prevent the development of resistance.

Clinical applications

Tuberculosis (as part of triple therapy); in combination with other antibiotics for endocarditis and osteomyelitis; as prophylaxis for close contacts of meningococcal and *H. influenzae* meningitis.

Nitroimidazoles

Metronidazole

Mechanism of action

Metabolized by nitroreductases to active intermediates which result in DNA breakages.

Pharmacokinetics

Well absorbed orally or *per rectum* and distributed into most tissues; metabolized in the liver and metabolites excreted in the urine.

Antimicrobial spectrum of activity

It is active against *Bacteroides* spp., other anaerobes, *Giardia lamblia*, *Trichomonas vaginalis* and other parasites.

Toxicity

Nausea, metallic taste; rarely peripheral neuropathy.

Resistance

Rare, but may occur as a result of reduced uptake or decreased nitroreductase activity.

Clinical applications

Treatment of anaerobic infections, giardiasis, trichomoniasis and amoebiasis.

Other antibiotics

Fucidin (fusidic acid)

- Well-absorbed; good penetration into bone and joints.
- Active against staphylococci.
- Acts on bacterial ribosome; resistance develops rapidly if used as monotherapy.
- Toxicity causes hepatic damage.
- Used in combination with flucloxacillin for serious staphylococcal infection (e.g. osteomyelitis, endocarditis).

Nitrofurantoin

- Well absorbed; excreted in urine.
- Used only in uncomplicated urinary tract infections because no significant blood levels are achieved.

Antimycobacterial agents

- Rifampicin (see above).
- Isoniazid: side effects include hepatitis and peripheral neuropathy.
- Pyrazinamide: may cause hepatic damage.
- Ethambutol: may cause optic neuritis.

Resistance to antimicrobial agents

Mechanisms

There are six main mechanisms by which bacteria become resistant to antimicrobial agents:

1 Alteration of the target site to reduce or eliminate binding of the drug to the target.

2 Destruction/inactivation of the antibiotic.

3 Blockage of transport of the agent into the cell.

4 Metabolic bypass, providing the cell with a replacement for the metabolic step inhibited by the drug.

5 Increased loss of drug from the cell by increased production of efflux pump.

6 Protection of the target site by a bacterial protein.

Alteration of the target site

This type of resistance is typical of antibiotics that act on the ribosome (e.g. erythromycin, fusidic acid and rifampicin), and of the quinolones which act on bacterial topoisomerases. Resistance occurs when a spontaneous bacterial mutant arises from within a previously sensitive bacterial population, and the presence of the antibiotic then allows the resistant mutant to be selected out.

The alteration of penicillin-binding proteins is a further example of bacteria altering the target site, resulting in the organism becoming resistant to β-lactam antibiotics (e.g. MRSA and penicillin-resistant pneumococci). Despite the considerable usage of β-lactam antibiotics, this type of resistance remains uncommon.

Destruction/inactivation of antibiotic

β-Lactamases

β-Lactamases cause hydrolysis of the β-lactam ring of penicillins and cephalosporins to form an inactive product. β-Lactamase production may be chromosomally or plasmid mediated.

Aminoglycoside-modifying enzymes

Production of aminoglycoside-modifying enzymes is normally plasmid mediated.

Interference with drug transport

Resistance to some antibiotics involves interference with the transport of these drugs into the bacterial cell, e.g., aminoglycosides and carbapenems.

Metabolic bypass

The antibiotic trimethoprim blocks folate metabolism by inhibiting the enzyme dihydrofolate reductase (see Figure 29.2). Bacterial strains resistant to trimethoprim synthesize a trimethoprim-insensitive dihydrofolate reductase (encoded on a plasmid), allowing the organism to bypass the action of the antibiotic. Sulphonamide resistance is mediated by a similar mechanism.

Efflux

Drug efflux may occur in a wide range of Gram-positive and Gram-negative bacteria. It is caused by increased expression of chromosomally encoded proteins, with an unknown normal function. Efflux tends to result in low level resistance, but may occur with other mechanisms to produce high-level, clinically relevant resistance. Antibiotics that may be effluxed include β-lactams, quinolones, tetracyclines, macrolides and aminoglycosides.

Target site protection

Access of an antibiotic to its target may be blocked by a bacterial protein. This is an uncommon mechanism of resistance. It causes resistance to tetracyclines and fluroquinolones. It is plasmid encoded.

The genetics of bacterial resistance

Resistance of bacteria to antibiotics is based on genetic changes, which allow the organism to avoid the action of the antimicrobial agent. Antibiotic-resistant genes may be passed between bacteria by a number of different vectors.

- Duplication of the bacterial chromosome takes place during cell division.
- Plasmids are extra-chromosomal pieces of DNA containing genes that confer on the bacteria a number of different properties. These genes are often non-essential, but allow the organism to survive under a variety of different conditions, giving the organism a survival advantage. Resistance to antibiotics is one of a number of properties that may be conferred on bacteria by plasmids.
- Transposons are small genetic elements consisting of individual or small groups of genes. These genetic elements have no ability to replicate, but can move from one replicating piece of DNA to another, e.g. from the bacterial chromosome to a

plasmid, and vice versa. Some transposons also behave like plasmids. Transposons play an important part in the evolution of bacterial resistance.

• Bacteriophages are viruses that infect bacteria and may act as a vector in carrying bacterial genes from one bacterial cell to another.

• Integrons consist of blocks of several antibiotic resistance genes; they usually include sulphonamide and aminoglycoside-resistance genes. New resistance genes may readily be added to an integron. Integrons are found in Gramnegative bacteria.

The existence of mechanisms for transferring genetic information (horizontal transmission) by plasmids, transposons and bacteriophages allows organisms to become antibiotic resistant.

Epidemiology of antibiotic resistance

The principle underlying the emergence and spread of antibiotic resistance among bacteria is that the prevalence of resistance is directly proportional to the amount of antibiotic used. This is illustrated by the increased antibiotic resistance found in countries with unrestricted use of antibiotics, in hospitals compared with the community, and in intensive care units compared with general wards. Problem areas in antibiotic prescribing include:

• the use of antibiotics without prescription
• the uncontrolled use of antibiotics in agriculture
• poor prescribing habits
• the absence of antibiotic policies, particularly in hospitals.

Principles of antimicrobial therapy

• The need for antibiotic therapy should be carefully considered; mild, self-limiting infections often do not require treatment.

• The choice of antimicrobial agent is dependent on a number of factors relating to the patient, the organism and the site of infection (Table 29.2).

• The route of administration, dose and length of treatment depend on the severity and type of infection. Intravenous antimicrobials are used for severe infections.

Table 29.2 Factors influencing choice of antimicrobial agent for treating infections.

CHOICE OF ANTIMICROBIAL AGENT
• Pharmacology
• Interaction with other drugs
• Penetration to site of infection
• Cost
• Toxicity
• Likely pathogens
• Antimicrobial spectrum of activity
• Age of patient
• Tolerance
• Underlying conditions (e.g. renal or hepatic failure)
• Patient allergies
• Pregnancy and breast-feeding

• Antibiotic combinations.

— antibiotics used in combination may be *additive* (combined effect equal to the sum of the individual agents), *synergistic* (combined effect greater than achieved with addition; e.g. gentamicin plus penicillin), or *antagonistic* (drugs inhibit the action of each other)

— the use of two (or more) antibiotics may broaden the spectrum of antimicrobial cover, e.g. cefuroxime and metronidazole may be used to treat abdominal sepsis to cover coliform organisms and anaerobes, respectively

— occasionally, antimicrobial agents are used in combination to prevent treatment failure and the development of resistance, e.g. triple therapy for TB

— some antibiotics contain a fixed combination of two agents, e.g. co-amoxiclav contains amoxicillin plus a β-lactamase inhibitor in the same preparation. The β-lactamase inhibitor protects amoxicillin from degradation by certain β-lactamase-producing bacteria.

• Antibiotic prophylaxis: antibiotics can be used for prophylaxis to prevent infections. Prophylactic antibiotics are used most frequently for preventing infection after certain types of surgery (e.g. abdominal); antibiotic prophylaxis for other conditions is discussed in Chapter 25.

Table 29.3 Adverse reactions of some antimicrobials.

ADVERSE REACTIONS	
Antimicrobial	**Adverse reaction**
Penicillins	Allergic reactions
Cephalosporins	Allergic reactions
Carbapenems	Allergic reactions
Aminoglycosides	Nephrotoxicity
	Ototoxicity
Vancomycin	Nephrotoxicity
	Ototoxicity
Sulphonamides	Folate deficiency
	Marrow depression
	Stephens–Johnson syndrome
Rifampicin	Hepatotoxicity
Isoniazid	Hepatotoxicity
Fucidin	Hepatotoxicity
Chloramphenicol	Aplastic anaemia

Adverse reactions and antibiotics

(Table 29.3)

Antibiotics may be associated with side effects common to all drugs, including allergic reactions and toxicity to various organs (e.g. hepatotoxicity, nephrotoxicity).

Specific complications associated with antibiotics relate to their depressive effect on normal commensal flora, e.g. oral candidiasis and *Clostridium difficile* colitis.

Laboratory aspects of antibiotic therapy

Assessment of bacterial sensitivity

A number of techniques are available to assess susceptibility of bacteria to antibiotics in vitro (Plate 50). These include nationally recognised tests such as the British Society for Antimicrobial Chemotherapy assay (Plate 56). In addition there are automated methods available (Plates 54 & 55).

- *Disc diffusion tests*: antibiotic-containing filter paper discs are placed on agar plates inoculated with the bacteria to be tested. After overnight incubation, susceptible bacteria will show a zone of growth inhibition around the disc; resistant bacteria grow up to the disc edge (Plates 50 & 56).
- *Minimum inhibitory concentration (MIC) assays*: these give a more detailed assessment of bacterial susceptibility. Suspensions of the test bacteria are incubated overnight with doubling dilutions of the antibiotic; the lowest concentration that inhibits growth is the MIC.
- *Minimum bactericidal concentration (MBC) assay*: this is similar to the MIC but the end-point is bacterial killing rather than growth inhibition.
- *Agar dilution susceptibility tests*: an antibiotic is incorporated into the medium in an agar plate at a set concentration. The test organism is applied to the plate in a standardized amount. Whether or not the organism grows will determine whether it is categorized as sensitive or resistant. Using plates containing a series of concentrations the MIC may be determined.
- *Growth rate-based susceptibility tests*: the test microorganism is placed in a container with a set concentration of antibiotic. The rate of growth of the microorganism is measured. Slow-growing microorganisms are sensitive to the antibiotic; faster-growing ones are resistant.

Therapeutic drug monitoring

This is required in patients receiving antimicrobials with a therapeutic threshold close to the toxic range; examples include aminoglycosides (e.g. gentamicin, tobramycin, amikacin) and glycopeptides (e.g. vancomycin). The results of serum monitoring are used to adjust subsequent dosing of the patient. Patients with renal or hepatic insufficiency may require monitoring of other antimicrobials.

Chapter 30

Antifungal agents

Selective action against fungi, without toxic side effects, is difficult to achieve and there are a limited number of useful antifungal agents. The principal ones used in clinical practice are the azoles, flucytosine, griseofulvin and amphotericin B.

Azoles

- *Mechanism of action*: azoles alter the fungal cell membrane by blocking biosynthesis of ergosterol resulting in leakage of cell contents.
- *Toxicity*: they may cause transient abnormalities of liver function; severe hepatotoxicity is a rare complication of ketoconazole therapy.

Fluconazole

- *Antifungal activity*: active against *Candida* species (some species including *C. glabrata* and *C. krusei* are resistant), *Cryptococcus neoformans* and *Histoplasma capsulatum*; ineffective against *Aspergillus* and *Mucor* species.
- *Pharmacokinetics*: oral and intravenous preparations are available; good penetration into various body sites. Fluconazole is excreted unchanged in urine. Adverse effects uncommon.
- *Clinical applications*: mucocutaneous and invasive candidiasis; cryptococcal infections, antifungal prophylaxis in immunocompromised patients (e.g. AIDS patients and transplant recipients).

Itraconazole

- *Antifungal activity*: active against *Candida*, spp., *Histoplasma capsulatum*, *Aspergillus* spp. and dermatophytes.
- *Pharmacokinetics*: only available orally but well absorbed. It is highly protein bound and degraded into a large number of inactive metabolites and excreted in bile.
- *Clinical applications*: dermatophytoses; mucocutaneous and invasive candidiasis; aspergillosis; histoplasmosis; blastomycosis.

Ketoconazole

Available as oral or topical forms and used principally as topical therapy for dermatophyte infections and cutaneous candidiasis. Not active against *Aspergillus* spp.

Miconazole

Available as topical, oral, and parenteral forms; used principally as topical therapy for dermatophyte infections and cutaneous candidiasis.

Voriconazole

- *Antifungal therapy*: active against *Aspergillus* spp. and *Candida* spp. and other mycoses.
- *Pharmacokinectic*: oral and intravenous prepara-

tions available. Cleared mainly by hepatic metabolism; has many drug interactions (e.g. warfarin, ciclosporins); generally well tolerated; can cause visual disturbances.

• Clinical applications: aspergillosis (superior to amphotericin).

Flucytosine (5-fluorocytosine)

• Synthetic fluorinated pyrimidine.

• *Mechanism of action*: deaminated in fungal cells to 5-fluorouracil which is incorporated into RNA in place of uracil, resulting in abnormal protein synthesis and inhibition of DNA synthesis. Resistance may arise during treatment.

• *Pharmacokinetics*: intravenous and oral forms available; good tissue and cerebrospinal fluid (CSF) penetration; excreted via the urine.

• *Antimicrobial spectrum of activity*: effective against most yeasts, including *Candida* (not *C. krusei*) and *Cryptococcus neoformans*. Resistance may occur during treatment, so used in combination therapy.

• *Toxicity*: marrow aplasia; need to monitor serum levels.

• *Clinical applications*: combined with amphotericin B in the treatment of serious cryptococcosis and systemic candidiasis.

Griseofulvin

• *Mechanism of action*: inhibition of nucleic acid synthesis and damage to cell wall by inhibiting chitin synthesis; resistance is rare.

• *Pharmacokinetics*: available in oral preparation only; well absorbed and concentrates in keratin, so useful in the treatment of dermatophytosis. Griseofulvin is metabolized in the liver and metabolites are excreted in the urine.

• *Antimicrobial spectrum of activity*: restricted to dermatophytes.

• *Toxicity and side effects*: generally well tolerated; urticarial rashes may occur.

• *Clinical applications*: dermatophytoses.

Amphotericin B

• *Mechanism of action*: damages the fungal cell membrane by binding to sterols; this results in leakage of cellular components and cell death.

• *Pharmacokinetics*:
— topical preparations (lozenges/mouth washes) for oral candidiasis
— parenteral preparation for intravenous administration; highly protein bound with low penetration into many body sites, including CSF; metabolized in the liver.

• *Antimicrobial activity*: broad spectrum of activity, including *Aspergillus*, *Candida*, *Blastomyces* and *Coccidioides*, spp., *Cryptococcus neoformans*, *Histoplasma capsulatum* and *Mucor* spp. Resistance is rare.

• *Toxicity*:
— anaphylactic reactions occur in some patients (a small test dose should be given before starting therapeutic doses); fever and chills common.
— renal tubular damage is a serious problem; renal function should be monitored regularly; doses should be increased slowly.
— liposomal amphotericin preparations (amphotericin complexed to microscopic lipid micelles) have been developed that allow higher doses of amphotericin to be used without toxicity.

• *Clinical applications*: amphotericin remains the most effective therapy for systemic mycoses, including disseminated candidiasis, cryptococcosis, aspergillosis and mucormycoses.

Echinocandin antifungal agents

• *Mechanism of action*: inhibit synthesis of cell wall, resulting in cell lysis and death.

• Includes caspofungin.

Caspofungin

• *Antifungal activity*: active against *Candida* and *Aspergillus* spp.; not active against *Cryptococcus neoformans*.

• *Pharmakinetics*: intravenous use only. Mainly metabolized in liver. Drug interactions include cyclosporin, rifampicin; side effects infrequent; not hepatotoxic or nephrotoxic.

• *Clinical application*: treatment of invasive candidiasis and aspergillosis.

Chapter 31

Antiviral agents

The development of new antiviral agents has increased significantly in the last few years. Several antiviral agents are currently licensed for treatment and many new drugs are being developed. Currently available drugs are primarily used for the treatment of human immunodeficiency virus (HIV), hepatitis B and C viruses, herpesviruses (herpes simplex virus or HSV, varicella-zoster virus or VZV, cytomegalovirus or CMV, and human herpesviruses HHV6 and -7), influenza virus and respiratory syncytial virus (RSV).

Mechanisms of action

There are three broad groups:

1 Those that directly inactivate intact viruses (viricidal) include ether, detergents, ultraviolet (UV) light; these are not used for treating patients (see Chapter 28).

2 Those that inhibit viral replication at cellular level (antivirals).

3 Those that modify host responses (immunomodulators).

Antiviral agents inhibit:

- virus attachment to cell receptors
- virus uncoating
- viral genome transcription and replication
- virus maturation
- virion assembly and release.

Immunomodulators may enhance any deficient host–immune response.

Drug resistance

Resistance to antiviral drugs has been documented to virtually all currently available compounds. Drug resistance results from point mutations or deletions within the viral genome. Most drug-resistant viruses are found in immunocompromised patients. Investigation for drug resistance has become an essential part of the clinical management of patients treated with antiviral compounds.

The main clinically valuable antiviral agents are presented below.

Anti-HIV compounds

Nucleoside reverse transcriptase inhibitors (NRTIs)

Examples include zidovudine (AZT), didanosine (ddI), zalcitabine (ddC), stavudine (d4T), lamivudine (3TC), abacavir, emtricitabine (FTC).

Structure

Thymidine analogues (AZT and d4T), cytidine analogues (ddC, 3TC and FTC), adenine analogue (ddI) and guanosine analogue (abacavir).

Mechanism of action

Phosphorylation by cellular enzymes to the triphosphate form. Inhibits viral RNA-dependent

DNA polymerase (reverse transcriptase RT). NRTIs act as chain terminators in the reverse transcriptase RT reaction.

Activity
HIV (type 1 and 2), hepatitis B (3TC and FTC only).

Resistance
This develops gradually, resulting in high-level resistant HIV mutants in treated patients

Clinical applications
HIV infection and post-exposure prophylaxis.

Route of administration
Orally.

Nucleotide reverse transcriptase inhibitor

The example of this is tenofovir.

Structure
Acyclic nucleoside phosphonate diester analogue of adenosine monophosphate,

Mechanism of action
Phosphorylation by cellular enzymes to the diphosphate form. Inhibits viral RNA-dependent DNA polymerase (reverse transcriptase). Acts as chain terminator in the RT reaction.

Activity
HIV-1 and -2, hepatitis B

Resistance
This develops gradually, resulting in high-level resistant HIV mutants in treated patients.

Clinical applications
HIV infection

Route of administration
Orally.

Non-nucleoside reverse transcriptase inhibitors (NNRTIs)

Examples include nevirapine, delavirdine and efavirenz.

Mechanism of action
The NNRTIs are non-competitive or mixed inhibitors of RT. They bind to a hydrophobic pocket on RT proximal to, yet separate from, the active site that is present only on binding of nucleic acid

Activity
HIV-1.

Resistance
This develops gradually, resulting in high-level resistant HIV mutants among treated patients.

Clinical applications
HIV infection.

Route of administration
Orally.

Protease inhibitors

Examples include saquinavir, ritonavir, indinavir, nelfinavir, amprenavir, lopinavir and atazanavir.

Structure
These compounds are oligopeptide analogues, very similar to the natural substrate of the HIV protease enzyme.

Mechanism of action
Bind to the active site of the protease enzyme and inhibit its activity.

Activity
HIV-1 and -2.

Resistance
This develops gradually, resulting in high-level resistant HIV mutants among treated patients.

Clinical applications
HIV infection

Route of administration
Orally.

Viral entry inhibitors

The example is enfuvirtide (T20).

Structure
This is a peptide made up of 36 amino acids corresponding to amino acid residues of the viral glycoprotein precursor gp160 and gp41.

Mechanism of action
Inhibits virus cell fusion through interaction with its homologous region in gp41.

Activity
HIV-1.

Resistance
This develops gradually, resulting in high-level resistant HIV mutants among treated patients.

Clinical applications
HIV infection.

Route of administration
Subcutaneous injection.

Anti-hepatitis B compounds

Lamivudine and Emtricitabine

See above as for anti-HIV compounds.

Adefovir

Structure
Acyclic phosphonate analogue of adenine monophosphate.

Mechanism of action
Phosphorylation by cellular enzymes to the diphosphate form. Inhibits viral RNA-dependent DNA polymerase (reverse transcriptase). Acts as chain terminator in the RT reaction.

Activity
Hepatitis B, HIV and other retroviruses and, to lesser extent, also herpesviruses.

Resistance
This develops gradually and less frequent than for lamivudine.

Clinical applications
HBV infection particularly for the treatment of lamivudine-resistant HBV infections.

Route of administration
Orally.

Anti-hepatitis C compounds

Interferon-alfa and pegylated interferon-alfa

Interferons are host cell proteins synthesized by eukaryotic cells that can inhibit growth of viruses. There are three major classes:α, β and γ.

Mechanism of action
Interferons have complex direct antiviral effects and enhance the immune response to viral infection.

Activity
Several viral infections but mainly hepatitis B and C, and papillomaviruses. Pegylated interferon is interferon-α attached to polyethelene glycol, to slow down the rate at which the body eliminates the molecule, enabling dosing to be less frequent.

Clinical applications
Chronic hepatitis C; other possible applications include papillomavirus infections.

Route of administration
Intramuscular injection.

Ribavirin

Structure
A synthetic triazole guanosine analogue.

Mechanism of action
Ribavirin interferes with viral mRNA production.

Activity
Ribavirin inhibits a wide range of RNA and DNA viruses, in particular paramyxoviruses (RSV), arenaviruses (Lassa, Junin) and adenoviruses.

Resistance
Not recognized.

Clinical applications
Orally in combination with interferon-α for the treatment of hepatitis C and as a small size droplet aerosol for the treatment of RSV infections in high-risk infants.

Route of administration
Orally.

Anti-herpesvirus compounds

Aciclovir and valaciclovir

Structure
A deoxyguanosine analogue. Valaciclovir is L-valine ester of aciclovir; it serves as a pro-drug of aciclovir and is better absorbed orally.

Mechanism of action
It is activated by viral thymidine phosphokinases and cellular kinases to the triphosphate form. It inhibits viral DNA polymerases and blocks viral DNA synthesis by acting as a chain terminator.

Activity
HSV-1 and -2, and VZV.

Resistance
This occurs with HSV and VZV. There are several mechanisms, including lack of viral thymidine kinase and altered DNA polymerase. Resistant viruses are detected mainly in immunocompromised patients.

Clinical applications
Mucosal, cutaneous, and systemic HSV-1 and -2 infections, i.e. keratitis, encephalitis, genital herpes, neonatal herpes and, for VZV infections, including chickenpox and shingles.

Route of administration
Aciclovir (orally and intravenous infusion), valaciclovir (orally).

Penciclovir and famciclovir

Structure
Penciclovir is a deoxyguanosine analogue. Famciclovir is a diacetyl ester of penciclovir and serves as a pro-drug of penciclovir; it has more bioavailabilty characteristics.

Mechanism of action
This is similar to aciclovir. It inhibits viral DNA synthesis.

Activity
HSV-1 and -2, and VZV.

Resistance
This can occur, mainly in immunocompromised patients.

Clinical applications
HSV-1 and -2 and VZV infections.

Route of administration
Orally.

Ganciclovir and valganciclovir

Structure
Ganciclovir is a deoxyguanosine analogue. Valganciclovir is the L-valine ester of ganciclovir and serves as a pro-drug of ganciclovir; it has a higher tissue concentration.

Mechanism of action
It is activated by viral HSV thymidine phosphoki-nases and CMV-encoded protein kinases and cellu-lar kinases to the triphosphate form. It inhibits viral DNA polymerases and blocks viral DNA syn-thesis by acting as a chain terminator.

Toxicity and side effects
Myelosuppression with: neutropenia (up to 40% of patients); thrombocytopenia (up to 20% of pa-tients); effects are reversible on cessation; need to monitor blood counts.

Activity
CMV, HHV6 and -7, HSV-1 and -2.

Resistance
Occurs mainly in immunocompromised patients.

Clinical applications
CMV infections, particularly in immunocompro-mised patients; prevention of CMV disease after organ transplantation.

Route of administration:
Ganciclovir (*intravenous infusion*), valganciclovir (*orally*).

Foscarnet

Structure
Foscarnet is an inorganic pyrophosphate analogue. It inhibits herpesviruses and most ganciclovir-resistant CMV, and aciclovir-resistant HSV and VZV mutants.

Mechanism of action
Foscarnet interferes with the binding of pyrophos-phate to its binding site of the viral DNA poly-merase and directly inhibits the enzyme.

Toxicity and side effects
Nephrotoxicity; hypo- and hypercalcaemia, and hypokalaemia.

Activity
HSV-1 and -2, VZV, CMV, and HHV6 and -7

Resistance
HSV, ZVZ and CMV can develop resistance, mainly in immunocompromised patients.

Clinical applications
CMV infections, aciclovir-resistant HSV and VZV infections.

Route of administration
Intravenous infusion.

Cidofovir

Structure
Cidofovir is a deoxycytosine analogue.

Mechanism of action
Phosphorylation by cellular enzymes to the diphosphate form. Inhibits viral DNA poly-merase and acts as a chain terminator of the viral DNA.

Toxicity
Nephrotoxic.

Activity
HSV-1 and -2, VZV, CMV, HHV6 and -7, and papil-loma-, polyoma-, adeno- and poxviruses.

Resistance
Resistance of herpesviruses mainly in immuno-compromised patients.

Clinical applications
CMV, HHV6 and -7 infections, aciclovir-resistant HSV and VZV infections, laryngeal and cutaneous papillomatous lesions, adenovirus infections.

Route of administration
Intravenous infusion.

Anti-influenza compounds

Amantadine/rimantadine

Structure
Tricyclic amines.

Mechanism of action
They inhibit an early step (uncoating) in the replication of influenza A virus.

Activity
Influenza A.

Resistance
Readily selected; up to 30% of patients have a resistant virus after treatment.

Clinical applications
Mainly treatment and prevention of influenza A virus infection.

Route of administration
Orally.

Neuraminidase Inhibitors (Zanamivir and Oseltamivir)

Structure
N-Acetylneuraminic acid (sialic acid) analogue.

Mechanism of action
Inhibit influenza viral neuraminidase activity and prevent viral release from infected cells. Viral neuraminidase is responsible for cleavage of N-acetylneuraminic acid from the influenza virus receptor and release of progeny virus particles.

Activity
Influenza A and B.

Resistance
Rare.

Clinical application
Treatment and prevention of influenza A and B infections.

Route of administration
Zanamivir is given by inhalation and oseltamivir orally.

Chapter 32

Upper respiratory tract infections

Infections of the upper respiratory tract are common and are primarily of viral origin and rarely serious.

Viral infections

Common cold

Aetiology
The majority of colds are caused by rhinoviruses. Other associated viruses are coronaviruses, influenza, parainfluenza viruses, respiratory syncytial virus (RSV), metapneumoviruses, adenoviruses and enterovirus. A number of systemic viral infections may present with similar upper respiratory tract symptoms, e.g. measles, mumps and rubella. Reinfections are common as a result of antigenic diversity within each of the viral groups.

Epidemiology
Transmission is by aerosol or large droplets or via virus-contaminated hands. Nasal discharge causes irritation and results in sneezing, facilitating spread.

Clinical features
The incubation period is 12 h to 2 days. Nasal discharge is often associated with cough, sneezing and non-specific symptoms, such as headache and malaise; fever is not a prominent feature. The nasal discharge may become purulent. Secondary bacterial infection by the pneumococcus, *Haemophila influenzae* or *Stretococcus pyogenes*, may result in sinusitis, otitis media or tracheobronchitis. Pharyngitis and conjunctivitis may occur with enterovirus or adenovirus infection.

Laboratory diagnosis
Virus isolation, direct antigen detection on nasopharyngeal aspirates or serological tests are usually not warranted because of expense and the benign course of the infection.

Treatment and prevention
There are no effective antiviral agents for treatment of the common cold viruses, except influenza; it is symptomatic treatment only. Hand washing helps prevent transmission.

Viral pharyngitis, tonsillitis and glandular fever syndrome

Aetiology
Many of the viruses associated with the common cold syndrome may also result in pharyngitis including rhinoviruses, coronaviruses, enteroviruses, adenoviruses, influenza and parainfluenza viruses and RSV. Epstein–Barr virus (EBV) and cytomegalovirus (CMV) cause glandular fever.

Epidemiology
Transmission is by aerosol and direct contact.

Clinical features

• Fever and pharyngeal discomfort with marked pharyngeal inflammation, occasionally with an exudate.

• The presence of vesicles on the soft palate indicates herpes simplex virus (HSV) or Coxsackie virus infection; conjunctivitis suggests influenza or adenovirus infection.

• Glandular fever syndrome: persistent pharyngitis with tiredness, malaise and cervical lymphadenopathy. This syndrome may be associated with an enlarged liver and/or spleen and jaundice. A maculopapular rash may occur after ampicillin or amoxicillin.

Laboratory diagnosis

Except in the case of glandular fever, laboratory diagnosis is rarely necessary. EBV and CMV infections may be diagnosed by specific serological assays.

Treatment

Symptomatic only.

Influenza

Aetiology

Influenza viruses A and B.

Epidemiology

Transmission is via aerosols or direct contact. Infection peaks during the winter months. Epidemics and pandemics of influenza A are related to varying degrees of antigenic variation.

Clinical features

Abrupt onset of fever, myalgia, pharyngitis and dry cough, which resolve within a week. However, tiredness may persist for several weeks. Complications of influenza are rare and include: primary viral pneumonia; secondary bacterial pneumonia (most frequently staphylococcal, now rare); encephalitis; Guillain–Barré syndrome; myocarditis and pericarditis; post-viral fatigue.

Laboratory diagnosis

Virus isolation, antigen detection in nasopharyngeal aspirates, polymerase chain reaction (PCR) or serology is not attempted routinely, but is important in confirming influenza epidemics.

Treatment and prevention

Treatment is for symptomatic relief only; the antiviral agent amantadine is active only against influenza A and is rarely used now. The neuraminidase inhibitors zanamivir and oseltamivir are active against influenza A and B. They may shorten symptoms if given within 48 h of onset and are usually given only to those at increased risk of developing severe complications: people with chronic cardiac and pulmonary disorders; patients with other chronic disorders (e.g. renal dysfunction, immunosuppression); and elderly people (> 65 years). Annual immunization before the start of the flu season is indicated for patients in these risk groups, together with healthcare workers.

Viral laryngotracheobronchiolitis (croup)

Aetiology

Parainfluenza viruses are the most frequent cause of croup; other viruses associated with upper respiratory tract infections can occasionally cause croup.

Epidemiology

Transmission is by aerosol spread. Most cases occur in the autumn and winter and the infection is restricted to children aged < 5 years.

Clinical features

Distinctive deep cough ('bovine cough'), frequently with inspiratory stridor, dyspnoea and, in severe cases, cyanosis.

Laboratory diagnosis

Virus isolation and antigen detection in nasopharyngeal aspirates (rarely performed).

Treatment

Symptomatic relief by mist therapy is often recommended, but there is no scientific proof of its effectiveness; antibiotics are of no benefit except when croup is complicated by bacterial infection. In severe cases, hospitalization is required and ventilatory support may be necessary.

Bacterial infections

Normal flora

Many of the bacteria found as part of the normal flora of the mouth are also present in the pharynx. Other commensal bacteria include neisseriae, *Haemophilus* spp. and *Streptococcus pneumoniae*. *Staphylococcus aureus* is carried in the anterior nares of about 30% of humans and may also be isolated from the pharynx. β-haemolytic streptococci colonize the upper respiratory tract of about 5% of individuals.

Streptococcal pharyngitis

Aetiology
Streptococcus pyogenes and occasionally other β-haemolytic streptococci (groups C and G).

Epidemiology
Approximately 5–10% of the population carry *S. pyogenes* in the pharynx; carriage rates are higher among children, particularly during the winter. Spread is by aerosol and direct contact, and is common within families; epidemics of streptococcal pharyngitis may occur occasionally in institutions, e.g. boarding schools.

Clinical features
Similar to viral pharyngitis; some features (e.g. systemic upset, purulent exudate) are claimed to be more prominent with streptococcal pharyngitis, but, in practice, it is difficult to separate viral from bacterial causes on clinical grounds alone.

Complications
• Local: peritonsillar abscess (quinsy), sinusitis, otitis media.
• Systemic: scarlet fever; rheumatic fever; acute glomerulonephritis, septicaemia and Henoch–Schönlein purpura.

Laboratory diagnosis
Throat swabs are taken for culture. Direct antigen

detection tests are available for rapid diagnosis. Serology (detection of anti-streptolysin O antibodies—'ASO titres') may be useful in patients presenting with post-streptococcal complications (rheumatic fever, glomerulonephritis).

Treatment
Penicillin, or erythromycin for penicillin-allergic patients; local analgesia—aspirin gargle.

Epiglottitis

Epiglottitis is an acute, severe infection of the epiglottis caused almost exclusively by *H. influenzae* type b. Since the introduction of the Hib vaccine, this infection has almost completely disappeared in the UK. Less common causes of epiglottitis include non-capsulated *H. influenzae*, *S. pneumoniae* and *S. pyogenes*.

Epidemiology
Before the Hib vaccine this infection was normally limited to children aged < 5 years, and spread by respiratory droplets. Adults may now rarely present with other causes of epiglottitis.

Clinical features
There is acute onset of fever, sore throat and respiratory distress with stridor. Lateral radiographs of the neck may show the enlarged epiglottis. Note that when the disease is suspected, direct visualization of the epiglottis should not be attempted, except where facilities for immediate intubation are available.

Laboratory diagnosis
Isolation of the organism from throat swabs or blood cultures.

Treatment and prevention
Intubation and ventilation may be necessary. Intravenous antibiotic therapy should be instituted immediately; second- or third-generation cephalosporins (e.g. cefuroxime, cefotaxime) are the current treatment of choice. Ampicillin should not be used as sole therapy, because 10% of strains of *H. influenzae* produce β-lactamase. Family con-

tacts of cases of *H. influenzae* b infection should be given rifampicin prophylaxis when there is a child in the family aged < 5 years old.

Vincent's angina

This is an uncommon throat infection caused by a mixture of anaerobic bacteria (e.g. fusobacteria, spirochaetes) normally found in the mouth. As with acute ulcerative gingivitis, with which it may coexist, it is associated with poor oral hygiene and underlying conditions such as immunodeficiency.

Clinical features include pharyngitis with a necrotic pharyngeal exudate. Microscopy of the exudate shows typical fusobacteria and spirochaetes. Treatment is with penicillin or metronidazole.

Diphtheria

Aetiology
An upper respiratory tract infection, caused by toxin-producing strains of *Corynebacterium diphtheriae*, with central nervous system and cardiac complications mediated by the exotoxin. Toxin is carried in pathogenic strains of *C. diphtheriae* in lysogenic bacteriophage. Occasionally skin infection may occur and result in the systemic effects of the toxin. Invasive infection has occurred in intravenous drug users.

Epidemiology
The organism is transmitted by aerosol spread. After the introduction of immunization, diphtheria is now rare in the developed world, but still common in developing countries.

Pathogenesis
The organism colonizes the pharynx, multiplies and produces toxin. The toxin inhibits protein synthesis. It acts locally to destroy epithelial cells and phagocytes, resulting in the formation of a prominent exudate, sometimes termed a 'false membrane'. Cervical lymph nodes become grossly enlarged ('bull-neck' appearance). The toxin also spreads via the lymphatics and blood, resulting in myocarditis and polyneuritis.

Clinical features
Fever and pharyngitis with false membrane (Plate 58) that may cause airway obstruction; enlarged cervical lymph nodes; myocarditis with cardiac failure; and polyneuritis. In skin infections chronic ulcers with a membrane form, and toxicity is mild. Such ulcers can act as reservoirs for respiratory infection and carriage.

Laboratory diagnosis
By isolation of the organism from throat swabs, followed by demonstration of toxin production by the Elek test, or more recently detection of toxin genes by PCR. PCR may also be used to detect toxin genes directly from throat swabs. Demonstrating toxin or the presence of the toxin gene is essential because non-toxin-producing strains of *C. diphtheriae* do not cause diphtheria.

Treatment and prevention
Patients with suspected diphtheria should be isolated in hospital. In suspected cases, treatment should be commenced before laboratory confirmation and includes antitoxin and antibiotics (penicillin or erythromycin). Diphtheria is a notifiable infection in the UK. Close contacts should be investigated for the carriage of the organism, given prophylactic antibiotics (erythromycin) and immunized. Childhood immunization with diphtheria toxoid has resulted in the virtual disappearance of diphtheria from developed countries.

Acute otitis media

Aetiology and pathogenesis
Upper respiratory tract infection may result in oedema and blockage of the eustachian tube with subsequent impaired drainage of middle-ear fluid, predisposing to viral or bacterial infection (acute otitis media). About 50% are caused by respiratory viruses; common bacterial causes include *Strep. pneumoniae*, *H. influenzae*, β-haemolytic streptococci and *S. aureus*.

Epidemiology
It occurs worldwide and is most common in

children aged < 5 years with an increased incidence in winter months.

Clinical features

Fever and earache, aural discharge. The eardrum appears reddened and bulging and, if untreated, drum perforations with subsequent purulent discharge may occur.

Laboratory diagnosis

By culture of discharge. Needle aspiration of middle ear fluid (tympanocentesis) is performed occasionally.

Management

Antibiotics prescribed include amoxicillin or erythromycin. Follow-up is important because residual fluid in the middle ear ('glue ear') can result in hearing impairment.

Otitis externa

Infections of the external auditory canal are frequently caused by *S. aureus* and *Pseudomonas aeruginosa*. Topical treatment with antibiotic-containing eardrops is often effective.

Acute sinusitis

Acute sinusitis follows impaired drainage of the sinus cavity and may complicate viral upper respiratory tract infections; it is also predisposed to by cystic fibrosis, nasal polyps, septal deviation and dental abscess. Bacterial causes are similar to otitis media. It presents with fever, facial pain and tenderness over affected sinuses; radiographs show fluid-filled sinuses. Management is with antibiotics (see 'Acute otitis media' above); occasionally sinus drainage is required.

Chapter 33

Lower respiratory tract infections

Pneumonia

Pneumonia is relatively common, with a significant morbidity and mortality.

Definition
It is an acute respiratory infection with focal chest signs and radiological changes. Various classifications are based on the causative organism, the radiological appearance and histopathological changes. A practical classification is:
- community-acquired pneumonia
- hospital-acquired pneumonia (nosocomial)
- pneumonia in immunocompromised individuals.

Community-acquired pneumonia

Epidemiology
This is an important infection worldwide. It is most common in the winter months; the overall incidence may vary in relation to outbreaks (e.g. *Legionella pneumophila*) or epidemics (e.g. *Mycoplasma pneumoniae*). Common viral causes of community-acquired pneumonia are shown in Table 33.1.

Clinical features
- *Symptoms and signs*: malaise, fever and shortness of breath; productive cough; pleuritic chest pain; tachypnoea and tachycardia; focal chest signs, e.g. consolidation; cyanosis.

- *Severe cases*: hypoxia, confusion, circulatory collapse resulting in renal and hepatic dysfunction.
- *Chest radiograph*: various appearances; lobar or patchy consolidation or diffuse shadowing.

Complications
Septicaemia and empyema (infected pleural effusion).

Laboratory diagnosis
- Sputum specimens for culture; microscopy (Gram stain) directly on sputum is of limited use in accurately predicting the causative organism.
- Bronchoalveolar lavage (BAL) specimens obtained by bronchoscopy for microscopy, culture and direct immunofluorescence tests.
- Blood cultures.
- Serology (viruses and *Mycoplasma*, *Chlamydia*, *Coxiella* and *Legionella* species).

Specific causes of community-acquired pneumonia

Streptococcus pneumoniae
This is a common cause of community-acquired pneumonia (30–50%) and occurs in all age groups. At-risk groups include patients with chronic lung disease, splenectomized patients and immunocompromised individuals, including those with HIV infection. Immunization is recommended for these groups. A chest radiograph commonly shows

Table 33.1 Common viral causes of pneumonia.

COMMON VIRAL CAUSES OF PNEUMONIA	
Virus	**Notes**
Influenza viruses	Epidemics/pandemics; elderly people and patients with chronic lung disease at risk
Parainfluenza viruses	Mainly in young children causing croup
RSV	Bronchiolitis in infants aged < 6 months
Measles virus	Rare; seen mainly in adults and immunocompromised patients
VZV	Rare; seen mainly in adults and immunocompromised patients
Adenoviruses	Young adults; associated with pharyngitis and conjunctivitis
CMV	Important in immunocompromised patients, particularly transplant recipients
HSV	Immunocompromised and neonates

lobar consolidation. Laboratory diagnosis is by sputum and blood culture. Treatment includes a penicillin or erythromycin, or a fluoroquinolone with anti-pneumococcal activity (levofloxacin or moxifloxacin). Complications include septicaemia, meningitis, pleural effusion and empyema.

Mycoplasma pneumoniae

Incidence varies (< 1–20%) as a result of epidemics; occurs primarily in young adults. Often presents with sore throat and dry cough, thus resembling influenza. The chest radiograph commonly shows diffuse changes. Laboratory diagnosis is by serology. Mild cases may resolve spontaneously. Treatment includes erythromycin or a tetracycline. Fluoroquinolones also have activity.

Haemophilus influenzae

Widespread use of the *H. influenzae* type b (Hib) vaccine has considerably reduced this cause of pneumonia in infants aged < 5 years. Most cases are now caused by strains other than Hib. Laboratory diagnosis is by sputum culture. Treatment is according to antibiotic susceptibility results using such antibiotics as trimethoprim, a tetracycline or fluoroquinolones.

Staphylococcus aureus

An important but now rare cause of post-influenza pneumonia; the incidence is highest in small children. Laboratory diagnosis is by sputum and blood culture. Treatment is with flucloxacillin.

Legionella pneumophila

L. pneumophila is transmitted via aerosols, particularly from contaminated air-conditioning systems or showers. Less than 5% of pneumonias are caused by *L. pneumophila* but incidence varies according to outbreaks. Hospital outbreaks may occur; cases associated with recent travel can arise. People who smoke or are elderly or immunocompromised are particularly at risk. The incubation period is 2–10 days. There is usually a dry cough at presentation. Laboratory diagnosis is by direct immunofluorescence on BAL specimens, sputum culture and serology. If infection is the result of *L. pneumophila* serogroup 1 (the most common cause of legionellosis), antigen may also be detected in urine. Treatment is with erythromycin or fluoroquinolones, with or without rifampicin.

Chlamydia pneumoniae

C. pneumoniae is another common cause of pneumonia in people aged 5–35. The symptoms are similar to those of mycoplasmal pneumonia. Laboratory diagnosis is by serology. Treatment is with erythromycin, a tetracycline or a fluoroquinolone.

Chlamydia psittaci

This is associated with contact with birds, including pigeons and parrots. Laboratory diagnosis is by serology. Treatment is with a tetracycline.

Coxiella burnetii

C. burnetii is a rare cause of pneumonia and is

usually confined to farm workers and vets. Laboratory diagnosis is by serology.

Viral pneumonia (see Table 33.1)

Primary viral pneumonia occurs mainly in children, elderly people and immunocompromised patients, with an increased incidence in winter. Underlying cardiopulmonary diseases are well-recognized risk factors for viral pneumonia in children and adults

Clinical manifestations in children vary considerably with the specific causative agent, but typically include fever, difficulty in breathing or apnoeic episodes in young infants, non-productive cough, wheezing or increased breath sounds. Common viral causes include: respiratory syncytial virus (RSV), parainfluenza viruses, influenza A and B, adenoviruses and measles.

Primary viral pneumonia in adults is characterized by non-productive cough, cyanosis and hypoxia, fever, rhinitis, increased respiratory rate, wheezes and diffuse bilateral interstitial infiltrates on chest radiographs. Most common causes include influenza, adenoviruses, varicella-zoster virus (VZV), RSV and severe acute respiratory syndrome (SARS).

Immunocompromised patients are at increased risk of severe viral pneumonia by viruses that are typical causes of lower respiratory tract disease in normal hosts and other more opportunistic viral pathogens (i.e. rhinoviruses). Cytomegalovirus (CMV) is the most frequent cause of pneumonitis in immunosuppressed patients particularly transplant recipients. Other common causes include herpes simplex virus (HSV), VZV, adenoviruses, RSV, and influenza A and B.

Diagnosis of viral pneumonia depends on detection of viral agents in respiratory specimens, i.e. nasopharyngeal aspirates, throat and nasal swabs and BALs. Viruses can be isolated in tissue culture and viral antigens can be detected by enzyme immunoassay (EIA) or by immunofluorescence techniques. Detection of viral nucleic acids by molecular techniques (i.e. polymerase chain reaction or PCR) is increasingly being used by diagnostic laboratories.

Specific antiviral therapy is available for some of the viruses associated with viral pneumonia (see Chapter 31). Ribavirin is used for the treatment of some cases of RSV and parainfluenza infections, and for adenovirus infections in immunocompromised patients. The neuraminidase inhibitors (oseltamivir and zanamivir) are available for the treatment of influenza A and B infections. Ganciclovir is frequently used for the treatment of CMV pneumonia whereas aciclovir and foscarnet are used for HSV and VZV infections.

Hospital-acquired (nosocomial) pneumonia

This is a lower respiratory tract infection (fever, chest signs, radiological changes) presenting 2 or more days after admission to hospital. Pneumonia is the third most common nosocomial infection, affecting about 0.5% of hospitalized patients. Risk factors include endotracheal intubation and ventilation, immunocompromise and pre-existing pulmonary disease. It is worth noting that, although patients may be in hospital, they may develop pneumonias from the same causes as community-acquired pneumonias.

Aetiology

Commonly *Streptococcus pneumoniae* and *Haemophilus influenzae* are causes, as well as Gram-negative organisms (ventilator-associated pneumonia or immunocompromise), such as *Escherichia coli*, *Klebsiella* and *Serratia* spp. and *Pseudomonas aeruginosa*. These organisms and anaerobes are also associated with aspiration of flora from the gastrointestinal tract, causing aspiration pneumonia. Meticillin-resistant *Staphylococcus aureus* (MRSA) is also well recognized as a cause of ventilator-associated pneumonia.

Pneumonia in immunocompromised patients

Immunocompromised patients may become infected with classic chest pathogens, e.g. *S. pneumoniae*, *M. pneumoniae*, or the important opportunistic lung pathogens (Table 33.2).

Table 33.2 Pathogens causing opportunistic chest infections in immunocompromised patients.

PATHOGENS CAUSING CHEST INFECTIONS	
Pathogen	Notes
Bacterial	
Actinomyceteae	Often cause abscesses and sinus formation
Atypical mycobacteria	Common in AIDS
Legionella pneumophila	May be hospital acquired
Fungi	
Candida species	Often preceded by oral candidiasis
Aspergillus and *Mucor* spp.	Transmission by spores; outbreaks have occurred in association with hospital building work
Pneumocystis jiroveci	Common in AIDS
Viruses	
CMV	Important in transplant recipients
HSV, VZV, measles virus	Important in immunocompromised children

Laboratory diagnosis

This is often difficult; early bronchoscopy for BAL fluid is important and should be examined by:

- microscopy (Gram and Ziehl–Neelsen stains)
- culture (bacteria, fungi, mycobacteria, viruses)
- direct immunofluorescence (*Pneumocystis jiroveci* and *Legionella* spp.).

Serology: atypical pneumonias, including *M. pneumoniae* and *L. pneumophila*.

Treatment

In the absence of clear aetiology, empirical therapy is often required. The choice depends on clinical presentation, laboratory data, underlying disease and previous antibiotic therapy.

Bronchiolitis

Bronchiolitis is caused by RSV and to lesser extent by parainfluenza and influenza viruses, and results in obstruction of bronchioles by mucosal oedema. Infection occurs mainly in infants aged < 18 months; it is particularly severe in infants aged < 6 months and with underlying cardiopulmonary diseases. Spread is by droplets. Clinical manifestations include fever, dyspnoea, respiratory distress and cyanosis. Laboratory diagnosis is by immuno-fluorescence or EIA for RSV antigen in nasopharyngeal aspirates, culture and PCR. Management may include nebulized ribavarin.

Acute exacerbation of bronchitis

This is principally viral (rhinoviruses, coronaviruses, influenza and parainfluenza viruses); rarely caused by *M. pneumoniae*. Clinical manifestations include dry cough, fever and malaise. Laboratory diagnosis is by culture or serology, but this is rarely performed.

Infective exacerbation of chronic obstructive pulmonary disease

Pathogenesis

The condition results in increased susceptibility to infection by *H. influenzae*, *S. pneumoniae*, *Moraxella catarrhalis* and some viruses.

Clinical features

Chronic sputum production (often mucoid but may be purulent) is part of the underlying condition. Acute-on-chronic infection occurs with increased production of purulent sputum and increasing dyspnoea.

Laboratory diagnosis

Sputum culture is performed for bacterial pathogens. Results need careful interpretation because causative agents are also commensals.

Treatment

Antibiotic treatment should be prescribed in relation to the clinical picture and sputum culture results, and includes amoxicillin, tetracycline and co-amoxiclav.

Bronchiectasis

There is underlying lung pathology with damage to the terminal bronchi and bronchioles, which act as a site of chronic infection. Pathogens include *H. influenzae, S. pneumoniae, Moraxella catarrhalis, P. aeruginosa* and anaerobes. Clinical features are similar to chronic bronchitis; antibiotic therapy is guided by sputum results and the patient's clinical condition.

Whooping cough (pertussis)

Aetiology and epidemiology

This is caused by *Bordetella pertussis*. Transmission is by aerosol. It is principally an infection of childhood but adults may occasionally be affected. Immunization has dramatically reduced the incidence in many countries.

Clinical features

The incubation period is 7–21 days, which has a prodromal phase (7–14 days) with coryzal symptoms. A severe cough develops; bouts of coughing are frequently followed by an inspiratory whoop. Complications include chronic lung disease and cerebral damage as a result of violent coughing (rare).

Laboratory diagnosis

This is by culture of perinasal swabs or 'cough plates'.

Treatment and prevention

Treatment is with erythromycin, which reduces infectivity if given early, but has little effect on the clinical course. The acellular component vaccine is given in childhood.

Tuberculosis

Epidemiology

Tuberculosis (TB) remains a common infection in the developing world, particularly in Asia and Africa; the current AIDS epidemic has led to an increase in cases. The incidence of TB in western Europe and North America has declined over the last 30 years as a result of improvements in nutrition, housing and various preventive measures, including Bacille Calmette–Guérin (BCG) immunization. More recently this decline has reached a plateau, and in some developed countries a small increase in cases has been reported. In the UK, about 6000 new cases, with about 500 associated deaths, are recognized each year; infection is more common in certain groups, including Asian immigrants and elderly men.

Pathogenesis

Primary TB: inhalation of *M. tuberculosis* results in a mild acute inflammatory reaction in the lung parenchyma, with phagocytosis of bacilli by alveolar macrophages. Bacilli survive and multiply within the macrophages and are carried to the hilar lymph nodes, which enlarge. The local lesion and enlarged lymph nodes are called the primary complex (referred to as the 'Ghon focus').

Histologically, granulomas consisting of epithelial cells and giant cells develop, which eventually undergo caseous necrosis. In many individuals, infection does not progress; however, organisms may remain viable for many years within lymph nodes. The local lesion and nodes become fibrotic and calcified.

In some patients, particularly immunocompromised individuals, organisms spread locally and via the bloodstream to other organs, causing widespread disease (miliary TB).

Secondary TB: may arise in two ways:

1 Dormant mycobacteria may become reactivated, often as a result of lowered immunity in the patient; reactivation occurs most commonly in the

lung apex, but may occur in other organs (e.g. kidney, bone).

2 A patient may become reinfected after further exposure to an exogenous source.

As with primary TB, local and distant dissemination may occur.

Clinical features
- *Pulmonary TB*: chronic cough, haemoptysis, weight loss, malaise and night sweats. Chest radiograph: cavitating lesion, frequently in apical regions, with enlarged hilar lymph nodes.
- *Extrapulmonary TB*:
  — *genitourinary*: sterile pyuria, with haematuria, pyrexia and malaise
  — *meningitis*: slow, insidious onset, with high mortality
  — *bone and joints*: rare, most commonly affects the lumbar spine
  — *lymph glands*: the most common site of non-pulmonary TB, the cervical lymph nodes are most frequently involved and may be the only site of infection, particularly in children
  — *abdominal TB*: rare, difficult to diagnose.

Laboratory diagnosis
Microscopy and culture of relevant specimens, depending on suspected site of infection, include sputum, bronchoscopy material, pleural fluid, urine, joint fluid, biopsy tissue and cerebrospinal fluid. Direct detection of *M. tuberculosis* DNA in specimen by the PCR is not generally recommended for primary diagnosis, because it has problems with sensitivity and specificity. Skin testing (Mantoux test) for diagnosis of active TB is appropriate only in countries where incidence of TB is low and BCG immunization is not administered routinely (e.g. the USA).

Treatment
This is often started on clinical suspicion; first-line agents include ethambutol, isoniazid, rifampicin and pyrazinamide. Combinations of up to four drugs are used to prevent emergence of resistance. As mycobacteria are very slow growing, treatment regimens range from 6 months to 1 year. Resistance to first-line agents is increasing particularly in Asia and Africa and, more recently, in areas of the USA, in association with HIV infection. Second-line antimycobacterial agents include ciprofloxacin, cycloserine amikacin, kanamycin and capreomycin.

Prevention and control
Strategies include: improved living standards (housing, nutrition); early recognition of new cases followed by isolation of patients with open TB for at least 2 weeks from start of therapy; follow-up of contacts, with skin testing and radiology as appropriate; prophylaxis with isoniazid for close contacts, particularly non-immune children and immunocompromised patients; immunization (BCG vaccine).

Chapter 34

Gastrointestinal infections

Definitions

Infective gastroenteritis

This is an inflammation of the gastrointestinal tract (GIT) caused by microorganisms. Spread is principally by the faecal–oral route, either directly or via vectors such as food or water. Symptoms include diarrhoea, vomiting, abdominal pain and fever.

Food poisoning and food-borne illness

This is the illness resulting from consumption of food contaminated with pathogenic organisms and/or their toxins. It mainly affects the GIT, but can affect other sites (e.g. *Listeria monocytogenes*—meningitis; *Clostridium botulinum*—paralysis). The incubation and clinical features of infections associated with food are summarized in Table 34.1.

It is difficult to distinguish with confidence, on clinical grounds alone, between microbiological aetiologies of infective gastroenteritis. However, knowledge of risk factors and incubation periods may hint at the aetiology, and are essential in public health management to prevent further cases. In the UK, statutory notification of food poisoning is made to the Consultant in Communicable Disease Control (CCDC) for the area of residence of the affected patient.

This chapter outlines the important causes of infective gastroenteritis and food poisoning. In addition, it deals with typhoid, which although acquired by the faecal–oral route, is not primarily an infective gastroenteritis. The key epidemiological and clinical features of the commonest infections are compared and summarized in Tables 34.2 and 34.3.

Host–parasite interactions

Infective gastroenteritis may result from the direct effect of organisms on the intestinal mucosa, or from the effect of toxins produced by bacteria.

Shigellae (see Chapter 10) invade the colonic mucosa, causing inflammation and ulcers, which result in mucosal bleeding. Diarrhoea is often bloody. Blood stream invasion is rare. *S. dysenteriae* also produces an exotoxin that disrupts absorption, resulting in diarrhoea.

Salmonellae (see Chapter 10) and campylobacters (see Chapter 13) multiply in the gut and cause direct mucosal damage; the overall increased fluid loss results in diarrhoea. Do not confuse *Salmonella* serotype *Typhi* or *Salmonella* serotype *Paratyphi* (which cause enteric fever) with the salmonellae that cause gastroenteritis. Enteric fever is primarily an infection of the monocyte–macrophage system with late GI symptoms from the involvement of Peyer's patches.

Vibrio cholerae (see Chapter 13) colonizes the mucosal surface of the intestine and produces an

Table 34.1 Some common bacterial causes and clinical features of food poisoning and gastroenteritis.

FOOD POISONING AND GASTROENTERITIS					
	Clinical features				
Organism	**Incubation period (h)**	**Diarrhoea**	**Vomiting**	**Abdominal pain**	**Fever**
Bacillus cereus	0.5–6.0	+M	+S	0	0
Campylobacter jejuni	48–120	+S	+M	+S	+
Clostridium perfringens	12–24	+M	0	+	0
Clostridium botulinum	12–36	0	±	0	0
Salmonellae	18–48	+M	±	+	0
Staphylococcus aureus	1–6	±	+S	0	0
Vibrio parahaemolyticus	6–36	+M	+M	+	0

0, absent; +, often present; ±, occasional; M, moderate; S, severe.

enterotoxin. Toxin stimulation of adenylyl cyclase in the mucosa causes intracellular cAMP (cyclic 3′,5′-adenosine monophosphate) to increase. This results in loss of water and electrolytes from the mucosa.

Antibiotic treatment leads to disturbance of the normal gut flora and allows *C. difficile* (see Chapter 7) to multiply. Some strains produce toxin A, which is responsible for the pathological change in the gut mucosa, and toxin B, a powerful cytotoxin that results in a cytopathic effect on tissue culture cells.

Although *Clostridium perfringens* (see Chapter 7) is part of the normal flora of the large intestine of animals and humans, it can cause food poisoning. Spores often contaminate meat and can survive cooking, particularly in large pieces of meat. If the food is then stored unchilled, the spores germinate and the organism multiplies. After ingestion, *C. perfringens* produces an enterotoxin (α-toxin) resulting in diarrhoea.

S. aureus is an important cause of toxin-mediated food poisoning. It contaminates food, typically via a food handler with a staphylococcal skin lesion. Contaminated food left at ambient temperature allows the organism to multiply and produce toxin, which can survive heating to 100°C for 30 min. Typical foods include cooked meat and cream cakes. After ingestion, the toxin is absorbed rapidly and acts on the central nervous system (CNS), resulting in severe vomiting.

The heat-resistant spores of *Bacillus cereus* are widespread and contaminate rice and other cereals. After surviving boiling of rice, the spores germinate if left at room temperature. The heat-stable toxin thus produced survives flash frying and causes rapid-onset (within 1–5 h) vomiting. A heat-labile toxin produced after ingestion can result in diarrhoea.

Although *E. coli* (see Chapter 10) is part of the normal flora of the large intestine, some of its strains may cause enteritis by a number of distinct pathogenic mechanisms:

1 Enteropathogenic *E. coli* (EPEC): this was an important cause of outbreaks of diarrhoeal disease among infants but is now much less common. It is caused by a number of distinct *E. coli* 'O' serotypes (e.g. O111, O127), which cause direct damage to the intestinal villi.

2 Enterotoxogenic *E. coli* (ETEC): ETEC is an important cause of travellers' diarrhoea and diarrhoeal disease in children in developing countries; ETEC produces enterotoxins similar to cholera toxin, which result in profuse, watery diarrhoea. Local people develop immunity to the ETEC strains to which they are exposed.

3 Verocytotoxin-producing *E. coli* (VTEC; see Table 34.3): VTEC causes two distinct conditions: haemorrhagic colitis (HC) and haemolytic–uraemic syndrome (HUS). VTEC strains produce a toxin similar to that of *Shigella dysenteriae* type 1. The toxin may act locally on the gut mucosa,

Table 34.2 Common gastrointestinal infections.

Infection	Shigella	Salmonella	Campylobacter	Clostridium difficile
Epidemiology	Humans are only host. Spread: faecal–oral. Very low infective dose (ca. 50 bacteria), thus person–person spread common. Food is occasional vector. Inanimate objects contaminated (e.g. toilet seats). In developing world in adults and children (S. dysenteriae and S. boydii). In developed world mostly young children (nurseries); S. sonnei and S. flexneri	Salmonella commonly carried in domestic and wild animals. Spread: faecal–oral via food (rarely water). Cross contamination of raw and cooked foods in kitchens. Direct person–person spread possible (hospitals). Long-term carriage may occur.	C. jejuni is most common species. Found in cattle, sheep, poultry, pets and wild birds. Spread: faecal–oral via food, but person–person spread possible. Contaminated poultry and unpasteurized milk. Long-term carriage uncommon.	Associated with use of antibiotics, especially cephalosporins and clindamycin. May be endogenous infection, or exogenous (person–person spread in hospital outbreaks).
Clinical presentation	Incubation: 1–4 days. Diarrhoea, with blood, mucus and pus. Fever rare. Symptoms last 3–4 days. S. dysenteriae associated with severe disease; S. sonnei via mild disease.	Incubation: 18–48 h. Diarrhoea, with abdominal pain and fever. Symptoms last 2–3 days.	Incubation: 2–5 days. Diarrhoea, with blood mucus and pus. Fever and abdominal pain prominent. May mimic appendicitis. Symptoms last 5–7 days.	May occur within 2 days of starting or up to 6 weeks after stopping antibiotics. Range of illness: mild disease to pseudomembranous colitis (severe bloody diarrhoea and systemic symptoms).
Complications	Dehydration and renal failure. Reactive arthritis associated with HLA-B27 (rare). Toxic dilatation and perforation; Haemolytic uraemic sydrome — S. dysenteriae.	Septicaemia (rare; in AIDS', rarely metastic spread, e.g. bone.	Guillain–Barré syndrome (rare).	Toxic dilatation and perforation Relapses.

Table 34.3 Some gastrointestinal infections associated with toxin-mediated disease.

Infection	Verocytotoxin-producing *E. coli*	*Vibrio cholerae*	*Staphylococcus aureus*	*Clostridium perfringens*
Epidemiology	Main reservoir is GIT of cattle. Very low infective dose (ca. 50 bacteria), thus person–person spread common. Prolonged excretion uncommon, but possible in children. Transmission by undercooked beef, unpasteurised milk, contact with farm animals. Commonest type is O157.	Humans are only host, but can survive in brackish water. Transmission usually by water and food contaminated by faeces. Person–person spread may occur. Prolonged excretion may last for several weeks. Epidemic cholera due to toxin-producing strains of serogroups O1 and O139.	*S. aureus* contaminates food, usually from a food handler with lesion. Typical foods are cooked meat and cream cakes.	Spores of (α-toxin-producing strains contaminate large pieces of red meat.
Clinical presentation	Incubation period 3–4 days. Spectrum from mild, non-bloody diarrhoea to haemorrhagic colitis. Fever rare. Symptoms usually end with within 1 week.	Incubation period 1–3 days Sudden onset of vomiting and copious watery diarrhoea (up to 25 L/day; 'rice water' stool). Fever absent. Usually resolves within several days.	Incubation period 1–6 h. Severe abdominal pain and vomiting. Diarrhoea is rare. Self-limiting—lasts 12–24 h.	Incubation period is 12–24 h. Abdominal pain and explosive diarrhoea. Self-limiting—lasts <6 h.
Complications	Haemolytic-uraemic syndrome • 2–14 after diarrhoea starts • children and elderly • risk of long-term renal failure Thrombocytopaenic purpura—adults.	Death from shock or renal failure. Death rates may be >50% in unprepared developing countries.	None	

resulting in bloody diarrhoea (HC), or act systemically, resulting in haemolysis (secondary to microvascular angiopathy) and renal failure (HUS).

4 Enteroinvasive *E. coli* (EIEC): EIEC produces a *Shigella*-like infection, with invasion of the intestinal mucosa, resulting in diarrhoea containing blood, pus and mucus.

Laboratory diagnosis

For many infections (salmonellae, shigellae, campylobacters, vibrios and *E. coli*), laboratory diagnosis is made by stool culture on selective media (see Chapter 24). The growth of an infecting organism allows it to be differentiated into a 'type'. Such typing allows the comparison of organisms and is useful in public health investigations. A rapid presumptive diagnosis of cholera may be made in the field in developing countries by using dark-field microscopy to recognize vibrios.

Diagnosis of *C. difficile* diarrhoea may be made by culturing the organism. Most commonly the diagnosis is made by immunological detection (enzyme-linked immunosorbent assay or ELISA) of toxins A and B in stool. Sigmoidoscopy (with biopsy) may demonstrate pseudomembranous colitis.

S. aureus food poisoning may be diagnosed by isolation of the organism from food and demonstration of its ability to produce enterotoxin. However, often no viable organisms are present in food. Direct detection of toxin in food can be made by immunoassays.

Management

Maintenance of hydration is essential for all patients with diarrhoea. Most diarrhoea in small children is viral—hence antibiotics are rarely ever necessary. In most instances, antibiotics would only slightly shorten symptoms and might select for resistance. Ciprofloxacin therapy may be considered for patients with continuing severe diarrhoea, from *Shigella*, *Salmonella* or *Campylobacter* spp., particularly if there are systemic symptoms such as fever, or for immunocompromised patients. However, many strains of these organisms have developed resistance to quinolones and other antibiotics. Antibiotics are indicated for invasive salmonellosis (ciprofloxacin), cholera (ciprofloxacin or tetracycline reduces duration of symptoms and excretion), or giardiasis and amoebiasis (metronidazole).

Antibiotics should not be used in the treatment of VTEC, as patients with this infection are more likely to develop its serious complications. This is because antibiotic exposure causes a 'stress' response in VTEC, which results in increased toxin.

Anti-motility drugs should be avoided, except perhaps for pragmatic reasons as a short course in travellers' diarrhoea when abroad. For similar reasons, ciprofloxacin might be considered for travellers' diarrhoea.

Prevention

Providing safe drinking water, proper disposal of sewage and food hygiene are major measures in preventing gastrointestinal infections. When patients are symptomatic, high standards of personal hygiene and providing hand-washing facilities are essential to prevent secondary cases. In hospitals, patients should be put in 'source isolation' until asymptomatic for 48h. Similarly, in the community, cases should be excluded from work or school until asymptomatic for 48h. Food handlers should have three negative consecutive stool samples at weekly intervals before being allowed to resume work.

In the UK, statutory notification of food poisoning, other causes of gastroenteritis and typhoid is made to the local Consultant in Communicable Disease Control (CCDC). The CCDC will coordinate the investigation of cases, and advise on the management of outbreaks and exclusion criteria for affected people.

Immunization does not have a major role to play in the prevention of gastrointestinal infection. Parenteral, killed, whole-cell cholera vaccines provide short-lived protection in only a proportion of recipients and are of little practical use. Oral cholera vaccines, suitable for use by travellers, have recently been developed.

Less common causes of food poisoning

Yersinia enterocolitica is associated with contaminated food, milk or water. The organism invades the terminal ileum and may result in a mesenteric adenitis, mimicking appendicitis. Bloodstream spread may occur in immunocompromised individuals or those with iron overload. This infection is associated with reactive arthritis.

Vibrio parahaemolyticus infection is associated with consumption of shellfish. Pain, fever and diarrhoea follow 12 h after ingestion.

Clostridium botulinum is a rare cause of food poisoning, more common in countries where there is small-scale canning of fruit and vegetables. *C. botulinum* spores contaminate the food and survive the heating process. The organism multiplies and toxin is produced. Pre-formed toxin is ingested and absorbed from the gastrointestinal tract, and it blocks neurotransmission of peripheral nerves, leading to a flaccid paralysis and eventually respiratory arrest. Symptoms usually appear within 12–36 h, occasionally several days, after eating contaminated food. Laboratory diagnosis is dependent on detecting toxin, in either food or faeces by culture or serological tests. Treatment involves intensive care with respiratory support, and administration of anti-toxin. Preventive measures in the canning industry involve heating food to temperatures high enough to destroy spores.

Viral causes of gastroenteritis

Viruses frequently result in gastroenteritis, particularly in children; in the developing countries, they are a major cause of death. Up to a million infants worldwide die of viral gastroenteritis every year. There is no specific treatment, except fluid replacement. In small children in developing countries, oral fluid replacement with electrolyte solutions is an extremely important part of management.

Laboratory diagnosis is by electron microscopy, by the direct detection of virus particles by immunoassays e.g. ELISA or PCR, depending on the virus being detected.

Rotaviruses

Epidemiology

Many different serotypes exist. Different serotypes cause diarrhoeal disease in other mammals, including cats, dogs, cattle, sheep and pigs. The infecting dose is small and spread is generally person to person. It is most common in children between 6 and 24 months old, but elderly people can be affected. Rotavirus infections are most common in the winter, with epidemics occurring occasionally in nurseries, elderly care homes and hospitals.

Host/parasite interaction

Damage to enterocyte transport mechanisms results in loss of water and electrolytes and diarrhoea.

Clinical features

The incubation period is 1–2 days. Vomiting and diarrhoea occur, often preceded by upper respiratory tract symptoms (cough, coryza). Infections are self-limiting and there is no specific treatment, apart from fluid replacement.

Noroviruses and saproviruses
(see Chapter 19)

Epidemiology

Causes 'winter vomiting disease'. Humans are the only reservoir of infection. The incubation period is 16–48 h. Both sporadic infections and outbreaks are common but clinical cases are usually diagnosed in outbreaks. Outbreaks occur year round but more frequently during the winter months in temperate climates. Outbreaks are associated with contaminated water supplies and food. They develop rapidly in institutions, including hospitals, to affect patients and staff. The vomiting caused by these viruses causes their spread directly by aerosol and indirectly by environmental contamination. The viruses are very stable in the environment. Control of outbreaks requires good hygiene and environmental cleaning.

Pathogenesis and clinical features

Infects mainly the villi of small intestine but the exact mechanism of diarrhoeal production is unknown. The hallmark of infection is the acute onset of vomiting and diarrhoea. Duration of illness from 12 to 60 h. Treatment is symptomatic.

Other viral causes

Adenoviruses (types 40 and 41)

The second most common cause of acute diarrhoea in young children. Incubation period is 8–10 days. Diarrhoea is milder but of longer duration compared with rotavirus gastroenteritis. Fever and vomiting are common. Transmission to family contact is rare. Diagnosis is by electron microscopy (EM), enzyme immunoassay (EIA) and PCR. Treatment is symptomatic.

Astroviruses

These viruses have a distinctive star-like surface appearance by electron microscopy. They have worldwide distribution. The peak incidence of infection is during the winter months in temperate climates. Transmission is by contaminated water or food. The incubation period is 24–36 h. It is a common cause of diarrhoea in infants and young children, and in elderly and institutionalized patients. Diarrhoea may be accompanied by fever, vomiting or abdominal pain. The illness lasts for 2–3 days. Diagnosis is by EM, EIA and PCR. Treatment is symptomatic.

Protozoal causes of gastroenteritis

Giardia lamblia

Epidemiology

G. lamblia has a worldwide distribution. It is spread by the faecal–oral route and is frequently associated with drinking water contaminated with cysts.

Host-parasite interactions

The life cycle of *Giardia* is described in Chapter 21. Trophozoites (Plate 40) attach to the mucosa of the

small intestine: epithelial cells are damaged, affecting transport mechanisms, impairing absorption and resulting in diarrhoea.

Clinical features

The incubation period is 1–3 weeks. Symptoms include diarrhoea (loose, fatty, foul-smelling stools), mild abdominal pain and discomfort, and are self-limiting (about 7–10 days). Chronic infection may occur in compromised hosts.

Laboratory diagnosis

By microscopy of stool samples or duodenal aspirates.

Treatment and prevention

Treatment is with metronidazole. Public health measures are important in prevention by ensuring clean drinking water.

Cryptosporidium parvum

Aetiology

Cryptosporidium is a well-recognized cause of diarrhoea in farm (especially calves) and wild animals. It also infects humans. The parasite's complex life cycle has asexual and sexual phases of development in the same host (see Chapter 21).

Epidemiology

Transmission is by the ingestion of cysts acquired either from direct contact with farm animals or infected cases, or via contaminated water or milk. The cyst releases sporozoites into the small intestine, which invade epithelial cells where they form schizonts and merozoites. These reinvade other epithelial cells. A sexual phase occurs with formation of oocysts.

Clinical features

Diarrhoea is normally mild and lasts several weeks. It can be severe in immunocompromised patients, particularly patients with acquired immune deficiency syndrome (AIDS), and continue for months or be fatal. An asymptomatic carrier state may occur.

Laboratory diagnosis

By microscopy of faecal samples stained by a modified acid-fast stain.

Treatment and prevention

There is no recognized effective treatment for cryptosporidial diarrhoea. Prevention includes care with personal hygiene when handling farm animals. As the cysts are not killed by chlorination, filtration is important in preventing contamination of drinking water supplies. Cysts will be killed if drinking water is boiled.

Entamoeba histolytica

Infections with *E. histolytica* are endemic in subtropical and tropical countries. The life cycle and associated infections are described in Chapter 21.

Typhoid and paratyphoid fever

These infections, although not primarily gastrointestinal infections, are acquired by the faecal–oral route. They are also known as enteric fever. Although the causative organisms, *Salmonella* serotype *Typhi* and *Salmonella* seotype *Paratyphi*, belong to the genus *Salmonella*, they are distinguished from other salmonellae by the nature of the infection that they cause. In addition, it is essential to differentiate them from the gastroenteritis-causing salmonellae.

Epidemiology

Enteric fever is prevalent in developing world. Paratyphoid is endemic in south-eastern Europe. In developed countries, enteric fever is mostly imported. Transmission is through food or water contaminated by faeces or urine. Person-to-personal spread is rare. Long-term carriage may occur. The incubation period is 10–21 days.

Host–parasite interactions

The organisms invade the intestine and spread to local lymph nodes, where they multiply. The infection enters a bacteraemic phase that coincides with the start of a fever. During this phase, the monocyte–macrophage system becomes infected (Peyer's patches, liver, spleen, bone marrow). The gallbladder is also involved. Inflammation and ulceration of Peyer's patches result in diarrhoea, and may lead to haemorrhage and perforation. After recovery, the bacteria may persist in the biliary and urinary tracts, resulting in chronic carriage. Abscesses in bone or the spleen may occur.

Clinical features

The first week of typhoid produces non-specific symptoms associated with the bacteraemia (fever, headache, constipation, relative bradycardia). It should therefore be included in the diagnosis of travellers returning from endemic areas with a fever. The second week of infection produces symptoms more associated with the GIT (diarrhoea, abdominal distension, 'rose spots' on the skin). Left untreated, the patient may have spontaneous recovery after a further 2 weeks or more. Paratyphoid is a much milder infection.

Laboratory investigation

Definitive diagnosis is from blood culture during the bacteraemic phase. Stool and urine cultures may be positive in the second week when the GIT is affected, but, because of the possibility of chronic carriage, must be interpreted in light of the clinical picture. The Widal test for 'O' and 'H' antibodies is too unreliable to contribute to diagnosis.

Treatment and prevention

Multi-drug resistance is present in south-east Asia. The use of ciprofloxacin (empirical and first choice) or ceftriaxone should be confirmed and guided by sensitivity tests. Amoxicillin, co-trimoxazole or chloramphenicol is a cheaper alternative when organisms are sensitive. Surgery may be needed for perforation. Chronic carriage may be cured by ciprofloxacin for 1 month. Typhoid and paratyphoid are notifiable infections in the UK. Three vaccines are available: killed whole cell (systemic side effects common), Vi capsular polysaccharide and the oral attenuated Ty21a vaccine.

Chapter 35

Liver and biliary tract infections

The liver has an important role in removing microorganisms and their products from the circulation. Patients with chronic liver disease also have lowered immune defences and are susceptible to infection.

The liver may become infected via the bloodstream (the dual blood supply of the liver via the hepatic artery and the portal vein results in the liver being particularly vulnerable to blood-borne infection) or via the common bile duct, particularly when bile drainage is obstructed. It may be involved in infection as a primary site (e.g. viral hepatitis) or as part of a multisystem infection (e.g. miliary tuberculosis, metastatic abscesses). Liver damage (jaundice, deranged liver enzymes) may be a feature of severe sepsis at other sites.

Cholecystitis is infection of the gallbladder; cholangitis infection of the bile duct. There is often an overlap between these two infections.

Aetiology and host–parasite interactions

Cholecystitis and cholangitis

The incidence of cholecystitis and cholangitis increases with age and is more frequent in women, as a result of an association with gallstones. Other factors predisposing to cholangitis are: biliary surgery or instrumentation (e.g. endoscopic retrograde pancreaticography or ERCP), biliary stricture or tumour, and pancreatitis. Obstruction of the bile duct results in stagnant bile, which becomes infected with commensal bacteria from the intestinal tract. Coliforms and enterococci are the most common organisms isolated; infections are frequently mixed and may include anaerobes (see Table 35.1).

Some parasites (see Chapter 22) are associated with biliary tract infection. *Fasciola hepatica* may cause biliary obstruction and cirrhosis. Schistosome eggs may lodge in the liver, resulting in liver fibrosis and subsequent portal hypertension.

Liver abscess — bacterial and amoebic

Bacterial liver abscesses may result from spread via the systemic circulation, the bile duct (cholangitis) or the portal vein (e.g. appendicitis, diverticulitis). There are multiple abscesses present in about a third of cases.

Amoebic abscess is a complication of intestinal amoebiasis caused by *Entamoeba histolytica* (see Chapter 21). Severe intestinal disease allows organisms to spread to the liver via the portal vein. The abscess increases in size and liver tissue is destroyed; amoebae are present at the margin of the abscess, the centre being filled with blood-stained pus. Abscesses are often single and in the right lobe of the liver.

The dog tapeworm, *Echinococcus granulosus*, may infect humans and result in hydatid cysts, particularly in the liver (see Chapter 22).

Table 35.1 Liver and biliary tract infections.

Infection	Cholecystitis	Cholangitis	Bacterial liver abscess	Amoebic liver abscess
Basic microbiology	Coliforms esp. *Escherichia coli*, *Klebsiella* and *Proteus* (see Chapter 10) Enterococci (see Chapter 6) Anaerobes, e.g. *Bacteroides fragilis* (see Chapter 15) Often mixed infection Rarely *Salmonella typhi* (see Chapter 10)		Additionally: *Streptococcus milleri* (Chapter 6) *Staphylococcus aureus* (Chapter 5) Rarely: *Mycobacterium tuberculosis* *Brucella sp.*	*Entamoeba histolytica* (Chapter 21)
Clinical presentation	Fever, right upper quadrant pain and tenderness, jaundice. Cholangitis usually associated with marked systemic toxicity. Palpable gallbladder may be present in cholecystitis. Greater increases in serum bilirubin and alkaline phosphatase in cholangitis than cholecystitis.		High, swinging fever with pain and tenderness in the right upper quadrant	Presentation up to 3 months after leaving endemic area for amoebiasis. Preceding dysentery not always present. Low-grade fever with right upper quadrant tenderness. Mild disturbance in liver function tests.
Investigation	Plain abdominal X-ray for calcification or gas in gallbladder or biliary tree Ultrasound scan of biliary tree Blood cultures (rarely positive in cholecystitis)		Imaging such as ultrasound and CT scans of liver Blood cultures (positive in 30% of bacterial abscesses; negative in amoebic abscess) Aspirated pus: microscopy for amoebic trophozoites; bacterial culture Serological tests (ELISA) for *E. histolytica* antibodies Stool for amoebic cysts and trophozoites	

Management	IV antibiotics to cover Gram-negative bacilli and anerobes (e.g. cefuroxime, or ciprofloxacin, or piperacillin, and metronidazole). Surgery: immediate only if complications; elective cholecystectomy after several months. ERCP & sphincterotomy for gallstones obstructing common bile duct.	Aspiration and percutaneous drainage under ultrasound guidance. Open surgery rarely needed	Empirical antibiotics (e.g ciprofloxacin or piperacillin and metronidazole) Modify after culture results	Metronidazole Diloxanide furoate to treat amoebic carriage
Complications	Perforation, gangrene, empyema of gallbladder Cholangitis Septicaemia	Perforation and empyema of gallbladder Liver abscess Pancreatitis Septicaemia	Subphrenic abscess, intrahepatic obstruction of major bile duct Rupture into surrounding structures Metastatic abscesses, e.g. lung, brain Septicaemia (bacterial abscess) Secondary bacterial infection of amoebic abscess	
Prevention	Treatment of gallstones	Prophylactic antibiotics for biliary surgery and ERCP Treatment of biliary tree obstruction	Prophylactic antibiotics for GI surgery	Food and water hygiene

Viral hepatitis

Viruses that primarily affect the liver are described as hepatotropic viruses. Currently five are recognized: hepatitis A, B, C, D and E. Viral hepatitis is a notifiable infection in the UK.

Hepatitis A virus (HAV)

HAV is an RNA virus and a member of the picornavirus family.

Epidemiology
• HAV is transmitted by the faecal–oral route. It is excreted in the faeces about 1 week before symptoms appear and up to 1 week after. Outbreaks have been associated with infected food handlers and ingestion of contaminated shellfish.
• Incidence is highest in the areas of the world with poor sanitation, and is most common in children.

Clinical features
• An incubation period of 2–6 weeks is followed by malaise, anorexia, nausea and right upper quadrant pain.
• Jaundice appears during the second week and normally lasts several weeks, but may be prolonged. About a quarter of patients are anicteric. Infections during childhood are generally asymptomatic but clinically apparent in infected adult patients. Fulminant liver failure occurs in about 0.1% of cases. Chronic liver disease is not a feature of HAV infection.

Laboratory diagnosis
• Liver function tests show markedly elevated serum levels of liver enzymes (aminotransferases) and bilirubin.
• Diagnosis is confirmed by the detection of serum IgM antibodies to HAV. Measurement of anti-HAV IgG is used to confirm immunity to HAV (past infection or immunization).

Treatment and prevention
There is no specific antiviral treatment; general measures include bedrest; hospitalization is rarely necessary. Passive immunization with normal immunoglobulin provides protection for about 6 months for travellers to endemic areas and for immediate protection of household contacts. A killed hepatitis A vaccine is now replacing the need for passive immunization.

Hepatitis E virus (HEV)

HEV is a small RNA virus, considered a calicivirus.

Epidemiology
Endemic in Indian subcontinent, south-east Asia, Middle East, north Africa and Central America. Spread by the faecal–oral route often via water; large waterborne epidemics have occurred in India. It is rare in the developed world except in travellers from endemic areas.

Clinical features
• The incubation period is 6–8 weeks, followed by mild hepatitis; severe fulminant HEV hepatitis may occur in pregnant women (10–20% of cases). Infections do not become chronic.

Diagnosis
Detection of IgG and IgM antibodies and by nucleic acid detection.

Treatment and prevention
Treatment is symptomatic. No vaccine available.

Hepatitis B virus (HBV)

HBV is a DNA virus and a member of the hepadnaviridae virus family.

Epidemiology
• Transmission of HBV is by percutaneous and permucosal routes. There are three important mechanisms of transmission:
— *contact transmission* via bodily secretions (e.g. blood, semen and vaginal fluid)
— *maternal–infant transmission* across the placenta or during delivery
— *percutaneous transmission*: at-risk groups include parenteral drug abusers and healthcare

workers involved in needlestick injuries. Infection via transfusion of contaminated blood products is rare in most countries because of active screening of blood donors.

• Prevalence patterns of HBV infection vary. In developed countries prevalence is < 1%, but in some endemic areas (e.g. north Africa, Asia) prevalence of the carrier state may be as high as 20%.

• In high-prevalence areas, maternal–infant transmission is important, whereas, in areas of low prevalence, percutaneous and sexual routes are more important modes of transmission.

Clinical features and complications

An incubation period of 6 weeks to 6 months is followed by malaise, anorexia and jaundice. Symptoms associated with immune complex disease occur rarely, e.g. arthralgia, vasculitis, glomerulonephritis. Asymptomatic infections in > 90% of infected children and 50% of infected adults. Complications include fulminant liver failure (1%) and chronic hepatitis (10%).

Laboratory diagnosis

Three HBV antigens and the corresponding antibodies are used in the diagnosis of acute and chronic HBV infection.

1 Hepatitis B surface antigen (HBsAg) is the first serological marker of acute HBV infection, appearing several weeks before symptoms.

2 Hepatitis B 'e' antigen (HBeAg) appears soon after HBsAg and is the first antigen to disappear in patients who are recovering from hepatitis B; its presence is associated with increased infectivity.

3 IgM antibodies to the hepatitis B core antigen are a useful marker of acute HBV infection in patients presenting after surface antigen is no longer detectable.

Presence of antibody to HBsAg, the last serological marker to appear, indicates past infection with HBV or previous immunization.

Treatment and prevention

There is no specific antiviral treatment for acute hepatitis B. Prevention is by: screening of blood donors and products, use of disposable needles and other instruments and efficient sterilization of re-

usable medical instruments. A recombinant HBsAg vaccine is available and should be given to at-risk groups, particularly healthcare workers. Specific immunoglobulin (passive immunization) may be given to non-immunized people exposed to HBV (e.g. in a needlestick injury) and to babies born to HBeAg-positive carrier mothers.

Chronic hepatitis B

About 10% of patients with acute hepatitis B will develop chronic disease. The incidence varies inversely with age; about 90% of neonates and < 10% of adults will develop chronic hepatitis B.

Definition

The persistence of HBsAg for > 6 months after acute infection.

Clinical features and outcome

About 80% of cases have minimal liver damage, remain largely symptom free and often undergo spontaneous remission (chronic persistent hepatitis). About 20% of cases follow a more aggressive course (chronic active hepatitis); liver damage is progressive and may result in cirrhosis. There is an increased risk of hepatocellular carcinoma.

Laboratory diagnosis

Based on failure to clear HBsAg. HBeAg may also persist and indicates continued viral replication and increased infectivity.

Management

Patients should reduce alcohol intake, which may act as a co-factor in the development of cirrhosis; liver function tests should be monitored. Treatment is with the antiviral agents lamivudine and adefovir. Liver transplantation may be performed in patients with end-stage liver failure.

Hepatitis C virus (HCV)

HCV is an RNA virus related to the pestivirus genus of the flavivirus family. There is high genome variability with at least six different genotypes (1–6) and several subtypes (1a, 1b, 2a, 2b, etc.).

Epidemiology

• HCV is the cause of about 90% of hepatitis cases previously known as 'non-A, non-B hepatitis' or 'non-HBV-transfusion-related hepatitis'.
• Transmission occurs in a similar manner to HBV (infected blood products, intravenous drug abuse, sexual transmission).

Clinical features

The incubation period is 2–6 months. Most infections are asymptomatic (80%) and in symptomatic cases the hepatitis is often mild. Some 90% of cases progress to chronic infection and if untreated can progresses to chronic active hepatitis and cirrhosis after many years (> 25 years), with an increased risk of developing liver carcinoma.

Laboratory diagnosis

This is by serology, for the presence of antibody to HCV and molecular techniques for the detection of HCV RNA.

Prevention

Methods are similar to those for HBV, including the screening of blood products. A vaccine has not been developed. Treatment is by interferon-α and ribavirin. HCV genotypes, in addition to other factors, can affect the response to treatment: genotype 1 is the least responsive.

Hepatitis D virus (HDV)

The hepatitis D virus or δ agent is a defective RNA virus that can replicate only in HBV-infected cells. Transmission is via infected blood and sexual intercourse. HDV accentuates HBV infection, resulting in more severe liver disease. Laboratory diagnosis is by HDV antibody (or rarely antigen) detection. Treatment and prevention are by treating HBV infections and the use of HBV vaccine.

Other causes of viral hepatitis

• EBV: in acute infectious mononucleosis. Usually mild a self-limiting infection but could be severe in some cases.
• CMV: acute or reactivation infections in immunocompetent or immunocompromised patients.
• Human herpesvirus 6 (HHV6): acute or reactivation infections in immunocompromised patients.
• HSV: in generalized infection in neonates or immunocompromised patients.
• VZV. As a complication of chickenpox or disseminated zoster in immunocompromised patients.
• Rubella: in congenital or acquired infections.
• Viral haemorrhagic fever viruses: yellow fever, rift valley fever virus, Ebola and Marburg viruses, Lassa fever virus, Junin and Machupo viruses.

Chapter 36

Urinary tract infections

Definitions

Clinical definitions

There are three clinical syndromes associated with urinary tract infection (UTI).

Frequency dysuria syndrome

Dysuria and urinary frequency caused by one of the following:

• Bacterial cystitis with significant bacteriuria (see below) and often associated with pyuria and haematuria due to acute inflammation. Also known as lower UTI.

• Abacterial cystitis ('urethral syndrome') — where no microbial cause is identified. Some cases may be associated with organisms that are not cultured by routine laboratory methods (Chapter 24).

Unlike urethritis caused by sexually transmitted infections, this syndrome is not associated with urethral discharge.

Acute bacterial pyelonephritis

Infection of the kidney with symptoms of loin pain and tenderness, and pyrexia accompanied by bacteriuria and pyuria. Also known as upper UTI.

Covert bacteriuria (asymptomatic bacteriuria)

This involves significant numbers of bacteria present in the urine of apparently healthy people, with no associated symptoms. It is important in childhood and pregnancy. Urinary catheterization is commonly associated with this condition.

Laboratory definition

Significant bacteriuria is defined as the presence of $\geq 10^8$ bacteria per litre (equivalent to $\geq 10^5$ bacteria/mL) of a fresh midstream urine (MSU) sample; occasionally patients with UTI symptoms may have fewer organisms (10^6–10^7/L), possibly as a result of dilution.

Aetiology and host–parasite interactions

Microorganism factors

The bacteria that commonly cause UTIs are commensals of the perineum or lower intestine (Tables 36.1 and 36.2). Non-bacterial causes of UTI are shown in Table 36.3. Bacterial virulence factors include (see also Chapter 23):

• Fimbriae: certain serotypes of *Escherichia coli* have specific fimbriae (pili) that facilitate colonization and adherence to the periurethral areas, urethra and bladder wall.

Table 36.1 Urinary tract infection.

Infection	Lower UTI Cystitis	Upper UTI Pyelonephritis	Catheter-associated UTI
Basic microbiology (see also Table 36.2)	*E. coli* (Chapter 10)—the most common cause *Proteus mirabilis* (Chapter 10)—associated with urinary tract stones Other enterobacteriaceae (e.g. *Klebsiella* spp. *Enterobacter* spp. and *Serratia* spp.; Chapter 10) associated with hospital-acquired UTI and catheterization *Pseudomonas aeruginosa* (Chapter 12)—associated with catheterization *Staphylococcus saprophyticus* (Chapter 5)—sexually active young women Non-bacterial causes: Viruses Adenoviruses (Chapter 19) Haemorrhagic cystitis Human polyoma virus (Chapter 19) Infections in kidney and ureter Parasites *Schistoma haematobium* (Chapter 22) Bladder inflammation; risk of malignancy in chronic infection		
Clinical presentation	Frequency of micturition Dysuria (pain on passing urine) Urgency Fever Non-specific symptoms (elderly & children): • poor weight gain, irritability (infants) • confusion, unsteadiness (elderly)	Loin pain or tenderness High fever • with symptoms of lower UTI	May present with non-specific symptoms and fever (especially elderly)

Laboratory diagnosis and result interpretation (see also Chapter 24)

MSU (midstream specimen of urine) or CSU (fresh specimen of urine via urinary catheter) for microscopy, culture and sensitivity testing (M, C & S) —to avoid perineal contamination. In infants use adhesive bag or suprapubic aspiration (SPA)

MSU (or CSU) and blood cultures (see Chapter 39)

CSU (catheter specimen of urine) for M, C & S. Fresh urine collected directly from catheter with syringe and needle (not from drainage bag)

Transport urine specimens to laboratory within 2 h to avoid bacterial growth and false diagnosis of UTI; if delayed, specimens should be refrigerated at 4°C or use the dipslide method (agar-coated slide dipped in urine, transported, then incubated).

Microscopy: for presence of leucocytes; in absence of catheter, $>10/mm^3$ indicates significant pyuria. Presence of erythrocytes and epithelial cells also quantified. Note neutropaenic patients may not produce pyuria.
Culture: various methods used to quantify number of bacteria
- $\geq 10^8$ bacteria/L ($\geq 10^5$ bacteria/mL)—significant bacteriuria
- $10^7–10^8$ bacteria/L ($10^4–10^5$ bacteria/mL)—may indicate infection; correlate with clinical symptoms, fluid intake or use of antibiotics that suppress bacteria growth
- $<10^7$ bacteria/L ($<10^4$ bacteria/mL) in patient not receiving antibiotics; infection unlikely—if patient is symptomatic, suggests abacterial cystitis.
Sample contamination more likely with two or more bacterial species grown; bacteriuria $<10^7$ bacteria/L; and/or presence of squamous cells from periurethra/perineum

Management

Increase intake of fluids to flush out infected urine

First line oral antibiotics in uncomplicated lower UTI: amoxicillin, trimethoprim, nitrofurantoin; short course 2–5 days. Persistent symptoms—repeat MSU 3 days after stopping antibiotics

Systemic antibiotics: e.g. ciprofloxacin, cefuroxime, gentamicin

May respond to first line antibiotics for lower UTI, but in context of hospital or previous treatment, there may be antibiotic resistance; be guided by sensitivity tests

Table 36.1 *continued*

	Lower UTI Cystitis	Upper UTI Pyelonephritis	Catheter-associated UTI
Infection			
Complications	Progression to pyelonephritis and/or septicaemia Epididymitis, prostatitis, chronic cystitis	Septicaemia Perinephric abscess Chronic pyelonephritis, scarring and renal failure	Progression to pyelonephritis and/or septicaemia
Prevention	Double micturition to ensure bladder is empty; micturition last thing at night. Good fluid intake	Treatment of cystitis	Aseptic technique when inserting catheter and maintaining drainage system
Special notes	Sterile pyuria — significant pyuria without apparent bacteriuria • consider renal tuberculosis; collection of three consecutive, complete early morning urine samples (EMUs) • other uncommon bacterial causes (e.g. mycoplasma) and non-bacterial causes (Table 36.3) • concurrent antibiotic treatment • non-infective causes: inflammatory renal disease, foreign bodies, tumours Asymptomatic bacteriuria in pregnant women needs treating to prevent pre-term delivery and low birth weight; use antibiotics that are not contra-indicated in pregnancy, e.g. beta-lactams Recurrent UTI may need investigation for anatomical abnormality (esp. children), tumours or diabetes		CSU: presence of bacteriuria and pyuria is common and does not always indicate infection

Table 36.2 Common bacterial causes of UTI (approximate percentage).

BACTERIAL CAUSES OF URINARY TRACT INFECTIONS		
Organism	Community acquired (%)	Hospital acquired (%)
Escherichia coli	75	40
Staphylococcus saprophyticus	5	1
Staphylococcus epidermidis	2	3
Proteus mirabilis	3	10
Enterococcus faecalis	5	8
Other coliforms	5	25
Pseudomonas aeruginosa	1	5
Candida albicans	4	8

Table 36.3 Non-bacterial causes of UTI.

NON-BACTERIAL CAUSES OF UTI		
Group	Organism	Infection
Viruses	Adenoviruses	Haemorrhagic cystitis
	Human polyoma virus	Infections in kidney and ureter
Parasites	*Trichomonas vaginalis*	Urethritis
	Schistoma haematobium	Bladder inflammation
Fungi	*Candida albicans*	UTI in immunocompromised patients

• Capsules: some strains of *E. coli* produce a polysaccharide capsule that inhibits phagocytosis and is associated with the development of pyelonephritis.

Host factors

Defence mechanisms against UTI include:
• Hydrodynamic forces: the flow of urine removes organisms from the bladder and urethra.
• Phagocytosis by polymorphs on the bladder surface.
• Presence of IgA antibody on the bladder wall.
• Mucin layer on the bladder wall prevents bacterial adherence.
• Urinary pH.

Given the importance of hydrodynamic forces in preventing UTI, the following are important host risk factors:
• Short urethra in females: sexual intercourse facil-itates the passage of microorganisms up the ure-thra (honeymoon cystitis).
• Structural abnormalities causing outflow ob-struction (e.g. prostatic enlargement, pregnancy, tumours); neurogenic bladder (e.g. in paraplegia); these result in residual urine in the bladder, which can act as a nidus of infection. Reflux of urine from the bladder up the ureter (vesicoureteral reflux) into the kidney can result from anatomical abnor-malities and cause pyelonephritis. Renal stones can be associated with pyelonephritis.
• Increasing age.

Other host risk factors:
• Diabetes mellitus; immunosuppression (e.g. steroids, cytotoxic drugs).
• Instrumentation (e.g. surgery or the use of uri-nary catheters); bacteria may be introduced into the bladder on catheter insertion, or may grow up the catheter into the bladder at a later stage. UTI associated with urinary catheterization is an important hospital-acquired infection.

Chapter 37

Genital infections (including sexually transmitted infections)

Most infections of the genital tract are spread by sexual contact; causative organisms are generally exclusive human pathogens and survive poorly outside the host. Sexually transmitted infections (STIs—Table 37.1) are an important cause of morbidity and mortality, and the incidence worldwide is increasing; in some developed countries the incidence of certain STIs (e.g. syphilis), has decreased because of active intervention programmes.

Gonorrhoea

Aetiology and epidemiology
- Caused by *Neisseria gonorrhoeae*, a Gram-negative diplococcus (Plate 18), an obligate pathogen of humans. Asymptomatic females act as a reservoir.
- The microorganism rapidly dies on drying, and direct contact is required for transmission; spread is via sexual contact or vertically from mother to neonate at birth. Gonorrhoea is an important STI worldwide; its incidence has recently increased in the UK.

Incubation period
Approximately 2 days

Clinical features
- In the male these are:
  — urethritis: purulent urethral discharge (Plate 59) and dysuria.
  — rectal: associated with homosexuals; most are

asymptomatic, but may cause proctitis (discharge and tenesmus).
  — pharyngitis: associated with oral–genital sex.
  — complications: prostatitis; epididymo-orchitis; bacteraemia with arthritis and skin lesions (e.g. pustules; Plate 60).
- In the female these are:
  — infection of the endocervix, urethra and rectum, symptoms (dysuria, vaginal discharge) may be mild or absent.
  — complications: bartholinitis (infection of Bartholin's glands); pelvic inflammatory disease (infection of salpinges with acute abdominal pain and pyrexia); bacteraemia with arthritis and skin lesions.
- Neonatal ophthalmia

Laboratory diagnosis
By Gram stain of smears for Gram-negative diplococci, often within pus cells (intracellular), and culture.

Management
- High-dose intramuscular procaine penicillin; β-lactamase-producing strains can be treated with spectinomycin or a cephalosporin (e.g. ceftriaxone).
- Contact tracing and investigation of sexual partners should be carried out.

Table 37.1 Common sexually transmitted infections

SEXUALLY TRANSMITTED INFECTIONS	
Organism	**Associated diseases**
Neisseria gonorrhoeae	Gonorrhoea
Human immunodeficiency virus (HIV)	AIDS
Chlamydia trachomatis	Non-specific urethritis
Candida albicans	Vaginal thrush, balanitis
Herpes simplex virus	Genital herpes
Treponema pallidum	Syphilis
Trichomonas vaginalis	Urethritis, vaginitis
Papillomaviruses	Genital warts

Non-specific urethritis

Definition
Urethritis in the absence of gonococcal infection.

Aetiology and epidemiology
This is a common STI mainly in males, with more than double the incidence of gonorrhoea. Causative organisms include *Chlamydia trachomatis*, *Ureaplasma urealyticum* and *Trichomonas vaginalis*; other organisms (e.g. group B streptococci, *Gardnerella vaginalis*, yeasts) may be associated with non-specific urethritis (NSU) occasionally. Mixed infections occur.

Clinical features
In male patients: mucopurulent urethral discharge, dysuria, and often difficult to distinguish clinically from gonorrhoea. In female patients: usually asymptomatic, but may present with urethritis or cervicitis.

Laboratory diagnosis and treatment
See individual infections below.

Chlamydial infections

Aetiology and epidemiology
C. trachomatis; serotypes D–K; transmitted during sexual intercourse. Vertical transmission to neonate at delivery.

Clinical features
Males: as in NSU. In females: cervicitis, NSU. May be asymptomatic in both males and females.

Complications
- Males: prostatitis, epididymitis, Reiter's syndrome (urethritis, arthritis, conjunctivitis).
- Females: bartholinitis, pelvic inflammatory disease.
- Neonates: conjunctivitis.

Laboratory diagnosis
This is by direct detection in urethral, cervical or urine samples, by ELISA (enzyme linked immunosorbent assay), by growth in tissue culture cell lines, and by polymerase chain reaction (PCR) assays.

Management
Tetracycline or erythromycin. Contact tracing and investigation of sexual partners should be carried out.

Trichomonas infection

Aetiology and epidemiology
T. vaginalis, a protozoan parasite, is transmitted during sexual intercourse.

Clinical features
- Female patients: vaginitis with copious, foul-smelling discharge.
- Male patients: up to 50% asymptomatic; they act as a reservoir of infection; may experience urethritis with urethral discharge and/or dysuria.

Laboratory diagnosis
Microscopy of a wet film of vaginal discharge shows motile trichomonads. Culture in special media can also be carried out.

Management
Treatment is with metronidazole; the sexual partner should also be treated.

Syphilis

Aetiology and epidemiology
This disease is caused by *Treponema pallidum*, a spirochaete and an obligate human pathogen. Syphilis has a worldwide distribution but is now uncommon in most developed countries. Transmission is by sexual contact and vertical spread via placenta to fetus.

Clinical features
Incubation is usually 14–21 days (range 10–90 days). Several clinical stages of untreated syphilis are recognized:
• *Primary syphilis*: a painless ulcer (chancre), develops on the genitalia, perianal area or, occasionally, at other sites. Treponemes multiply at the chancre and in the regional lymph nodes, which become enlarged. Lesions heal over several months.
• *Secondary syphilis*: this occurs up to 2 months after the first stage. It is associated with: pyrexia, malaise, sore throat; a widespread maculopapular rash and lymphadenopathy; wart-like lesions (condylomata lata) may occur in perianal, vulva or scrotal areas; and oral and pharyngeal ulcers.
• Lesions heal and a dormant phase follows, *latent syphilis*, which may last up to 30 years before tertiary syphilis develops. Dormant treponemes may be present in the liver and spleen.
• *Tertiary syphilis*: treponemes start to multiply, resulting in granulomatous lesions (gummas) in various sites, including skin, subcutaneous tissues, mucous membranes, bones and joints. Cardiovascular complications include aortic valve incompetence, aortic aneurysm; central nervous system (CNS) complications (neurosyphilis) include asymptomatic meningitis, tabes dorsalis (spinal cord damage) and general paralysis of the insane (cerebral damage).
• *Congenital syphilis*: spread via transplacental route; may result in intrauterine death or disease, presenting with congenital malformations at birth or several years later. Features include skin lesions, lymphadenopathy, failure to thrive, mental deficiency, peg-shaped teeth (Hutchinson's incisors), and bone and cartilage destruction (e.g. saddle-shaped nose).

Laboratory diagnosis
Dark-ground microscopy of exudates collected from lesions; a combination of several serological tests.

Treatment and prevention
Treatment is with penicillin or doxycycline in penicillin-allergic patients. Prevention depends on adequate treatment, safe sex, contact tracing, screening and treatment of sexual partners. Congenital syphilis is prevented by serological screening in early pregnancy.

Non-specific vaginosis (NSV)

NSV is a vaginal infection with no specific aetiological agent. Characteristic features of NSV include: watery, foul-smelling ('fishy') discharge (no pus cells); and absence of pruritus and pain. Microscopy of discharge demonstrates the presence of large numbers of Gram-negative rods, often adherent to vaginal epithelial cells ('clue cells'; see Plate 23).

A number of bacteria are associated with NSV, including *Gardnerella vaginalis* and *Prevotella* (see Plate 25). Treatment is with metronidazole.

Candidiasis

Candida albicans causes vulvovaginitis (vaginal thrush) in female patients. Low numbers of the organism are present as part of the vaginal commensal flora in some women; factors such as antibiotic therapy, steroid therapy (including the contraceptive pill), pregnancy and immunosuppression may disturb the normal flora and result in overgrowth of *C. albicans*. Clinical features include pruritus and a creamy vaginal discharge. Male patients become infected via sexual contact with infected females; the resulting infection involves the glans (balanitis), with inflammation and white plaques.

Laboratory diagnosis is by microscopy and culture of appropriate specimens. Treatment includes topical antifungals (e.g. nystatin) or oral fluconazole.

Granuloma inguinale

This genital infection is caused by *Calymmatobacterium granulomatis* and is usually found in the tropics. Sexual transmission may play a role in spread. Clinical features include papules and ulcers on the genitalia and regional lymph node involvement. Diagnosis is by microscopy of stained smears or biopsies of lesions. Treatment is with doxycycline.

Chancroid

This is a genital infection characterized by painful ulcers with lymphadenopathy and caused by *Haemophilus ducreyi*. It has a worldwide distribution but occurs principally in tropical countries.

Diagnosis is by microscopy of Gram-stained smears of aspirates of ulcer or lymph nodes. Culture requires enriched media and prolonged incubation. Treatment is with erythromycin or ciprofloxacin.

Lymphogranuloma venereum

This infection is characterized by a genital ulcer with inguinal lymph node enlargement and suppuration. It is caused by *Chlamydia trachomatis* (serotypes L1, L2 and L3) and has a worldwide distribution, but is more common in tropical countries.

Dissemination of the organism may result in complications, e.g. proctitis, hepatitis, arthritis and meningoencephalitis. Chronic infection may result in genital strictures and fibrosis of the lymphatics, with subsequent blockage and genital elephantiasis.

Diagnosis is by serology and treatment is with doxycycline or erythromycin.

Genital herpes

Aetiology and epidemiology
An STI caused generally by herpes simplex virus (HSV) type 2, although up to 40% of cases are caused by type 1 strains. Both viruses have worldwide distribution. HSV1 acquired early in life with > 90% of general population seropositive by adulthood. HSV type 2 infections mainly acquired in adolescence and early adulthood and seroprevalence rates are generally affected by social group and sexual activity (up to 90% seroprevalence among female commercial sex workers).

Clinical features
- Primary infection has an incubation period of 3–7 days. Clusters of vesicles appear on the genitalia, progressing to painful, shallow ulcers. There is local lymphadenopathy, fever, dysuria associated with urethritis and cystitis. Complications include aseptic meningitis and sacral radiculomyelitis with urinary retention.
- Resolution occurs in around 10 days, but HSV becomes latent in dorsal root ganglia; various stimuli (e.g. hormonal changes) may result in virus reactivation and recurrent lesions (genital cold sores). Recurrent infections are less severe with limited number of vesicles, less pain and rare complications.
- In pregnancy, active HSV infection at term, particularly primary infections, may result in transmission to the neonate, with invasive HSV infection (e.g. encephalitis).

Laboratory diagnosis
Vesicle fluid or ulcer swabs are taken for virus isolation or direct detection of virus by immunofluorescence or molecular techniques.

Treatment
Systemic aciclovir, valaciclovir and famciclovir can reduce symptoms and infectivity.

Genital warts

Aetiology and epidemiology
This is an STI caused by human papillomaviruses (HPVs). This family of viruses includes more than 100 types (see Chapter 19). Between 40 and 50 of these types are associated with genital infections, the most common being types 6, 11, 16 and 18. The viruses have worldwide distribution. Genital infections are acquired mainly during

adolescence and early adulthood through sexual intercourse.

Clinical features

After an incubation period of several months, warts appear on the external genitalia (penile shaft, labia minora and majora, anal region) and also at mucosal surfaces (vagina and cervix). Infections with high-risk types (types 16, 18 and 33) are associated with cervical carcinoma. HPV DNA can be detected in 90–95% of cervical cancer lesions. HPVs are also associated with anal and penile carcinoma.

Laboratory diagnosis

Cytology, serology, nucleic acid detection.

Treatment

Surgery, cauterization, cryotherapy and laser therapy, interferon, imiquimod (podophyllum resin).

Human immunodeficiency virus and AIDS

See Chapter 47.

Chapter 38

Infections of the central nervous system

- *Meningitis* is defined as infection of the meninges, the membranous covering of the brain and spinal cord.
- *Encephalitis* involves infection of the brain tissue, but is frequently accompanied by inflammation of the meninges (meningoencephalitis).
- Infection may result from haematogenous spread (e.g. *Neisseria meningitidis*), invasion via nerves (e.g. herpes simplex virus) or direct spread from a local focus of infection (e.g. mastoiditis).
- Infections of the central nervous system (CNS) may result in long-term sequelae (e.g. cranial nerve damage, epilepsy, deafness) as inflammatory changes and/or raised intracranial pressure can cause irreversible damage to neuronal tissue.

Bacterial meningitis

The principal causes of bacterial meningitis are shown in Table 38.1.

Major causes of bacterial meningitis

Neisseria meningitidis

- Upper respiratory tract colonization occurs in up to 10% of healthy individuals; this may increase during epidemics. Spread is by the respiratory route (droplets). Invasion of the meninges follows bacteraemia/septicaemia.
- There is a small increase in risk of infection among close contacts of patients with meningococcal meningitis.
- Infections occur principally in children and young adults. In the USA and Europe, cases of meningococcal meningitis are sporadic and caused mainly by serotype B; epidemics of *N. meningitidis* serotypes A and C occur in Africa and South America.

Clinical features
- Normally of rapid onset, the principal symptoms are headache, fever and meningism (neck stiffness, irritability, photophobia). Concurrent septicaemia frequently leads to a typical petechial rash, which may become haemorrhagic.
- In fulminant cases, endotoxin release results in shock, with disseminated intravascular coagulopathy and multi-organ failure. In a few cases, haemorrhagic necrosis of the adrenal glands and intracranial bleeding occur, which is invariably fatal (Waterhouse–Friderichsen syndrome).

Prophylaxis and prevention
- In the developed world epidemics are rare; however, related cases in families or institutions can occur and chemoprophylaxis with rifampicin or ciprofloxacin is recommended for close contacts.
- Healthcare workers in contact with cases are not at increased risk, but chemoprophylaxis is recommended for those involved in mouth-to-mouth resuscitation.

Table 38.1 Causative organisms of acute bacterial meningitis according to age group.

BACTERIAL MENINGITIS	
Age	**Causative microorganism**
< 1 month	β-Haemolytic streptococcus group B, *E. coli*, *Listeria monocytogenes*
1–23 months	β-Haemolytic streptococcus group B, *E. coli*, *Haemophilus influenzae*
	Streptococcus pneumoniae
2–50 years	*S. pneumoniae*, *Neisseria meningitidis*
≥ 50 years	*S. pneumoniae*, *N. meningitidis*, *L. monocytogenes*
	Aerobic Gram-negative bacilli—coliforms

• A vaccine is available for protection against serotypes A and C, and is recommended for prophylaxis of close contacts of meningococcal cases caused by these serotypes.

Treatment

This is with intravenous penicillin; cefotaxime or chloramphenicol may be used in penicillin-allergic patients. Rifampicin or ciprofloxacin is used to clear nasopharyngeal carriage.

Haemophilus influenzae type b

• This occurs principally in children aged < 5 years. Spread is by the respiratory route and symptomless upper respiratory colonization is common.
• Cases are often sporadic, but small outbreaks, particularly in nurseries, may occur.
• Number of cases have reduced as a result of introduction of *H. influenzae* type b vaccine.

Clinical features

It is similar to meningococcal meningitis except that onset is often less acute (1–2 days) and, although a petechial rash can occur rarely, it is much less common.

Treatment and prophylaxis

Antibiotic therapy with cefotaxime and/or chloramphenicol. Rifampicin is used for chemoprophylaxis of close contacts. A vaccine is available and should be considered for household contacts.

Streptococcus pneumoniae

• Infection is most common in young (< 2 years of age) and elderly individuals; asplenic patients (e.g. with sickle cell disease) are at particular risk.
• Meningitis may be secondary to pneumococcal septicaemia or skull fracture (direct spread).
• Pneumococci are common commensals of the upper respiratory tract. Person-to-person spread is not a feature of pneumococcal meningitis.

Clinical features

Similar to other causes, but *S. pneumoniae* infection may be acute or insidious in presentation. Complications (e.g. deafness) are more common than with other bacterial causes.

Treatment

Until recently, pneumococci have remained sensitive to penicillin and erythromycin. However, penicillin-resistant strains have now emerged and are a particular problem in South Africa and parts of Europe. In the UK, up to 10% are resistant to penicillin. In these cases, cefotaxime is often used.

Prophylaxis and prevention

Contacts of cases of pneumococcal meningitis are not at increased risk and chemoprophylaxis is inappropriate. A vaccine against many of the pneumococcal serotypes is available and is recommended for splenectomized and other immunocompromised patients.

Laboratory diagnosis of bacterial meningitis

Meningitis is a medical emergency and laboratory investigations are urgent.

- A lumbar puncture (LP) for cerebrospinal fluid (CSF) is carried out after raised intracranial pressure has been excluded. The typical features of CSF microscopy in bacterial meningitis are a raised polymorph count, a low glucose (compared with blood glucose) and an increased protein content (Table 38.2). Gram stain result may indicate the causative organism; culture is important for confirmation.
- Assays for *N. meningitidis*, *H. influenzae* type b and *S. pneumoniae* antigens in CSF and urine can provide an early diagnosis, and may be useful in patients who have received antibiotics before LP.
- Blood cultures are positive in about 40% of cases of bacterial meningitis.
- Serology may give a retrospective diagnosis of meningococcal meningitis.
- Polymerase chain reaction (PCR) tests for detection of causative organisms are becoming more widely available.

Treatment of bacterial meningitis

The fundamental principle in the treatment of bacterial meningitis is early, parenteral antibiotics in high doses. Intramuscular penicillin administered early by primary care physicians has been shown to reduce morbidity and mortality from meningococcal meningitis. Antibiotic therapy for common causes of bacterial meningitis in children and adults is shown in Table 38.3. There is some evidence that concomitant steroid therapy, when commenced at the same time as antimicrobial therapy, reduces morbidity and mortality.

Neonatal meningitis

Aetiology

Escherichia coli, group B β-haemolytic streptococci and *Listeria monocytogenes* are important causes of meningitis in the neonatal period; the three main causes of meningitis in children and adults are uncommon during the neonatal period.

Epidemiology

Acquisition is normally from the maternal genital or alimentary tract at or around the time of delivery.

- *E. coli* is a common gut commensal and frequently contaminates the perineal area.
- Group B streptococci are part of the vaginal and perineal flora of about 30% of mothers.
- *L. monocytogenes* is a gut commensal in a small proportion of healthy mothers.

Neonates with prolonged hospital stay may acquire these organisms from nosocomial sources. Prematurity, low birthweight and prolonged ruptured membranes are important risk factors for neonatal meningitis.

Table 38.2 Typical CSF changes in meningitis.

CEREBROSPINAL FLUID CHANGES				
		Meningitis		
	Normal	Bacterial	Tuberculous	Viral
Leukocytes (cells/µL)	<5	500–20000	200–1000	50–500
Polymorphs (%)		90	20[a]	5[a]
Lymphocytes (%)		10	80	95
Protein (g/L)	0.2–0.45	↑↑	↑↑↑	↑
Glucose (mmol/L)	4.5–8.5[b]	↓↓	↓↓	Normal

[a] Polymorphs may predominate in the early stages.
[b] About 80% of blood glucose level.

Table 38.3 Antimicrobial therapy for common causes of bacterial meningitis.

ANTIMICROBIAL THERAPY OF ACUTE BACTERIAL MENINGITIS	
Organism	Treatment
Aetiology unknown[a]	Benzylpenicillin; or cefotaxime or chloramphenicol
N. meningitidis	Benzylpenicillin (chloramphenicol or cefotaxime if penicillin allergic)
H. influenzae type b	Cefotaxime or chloramphenicol
S. pneumoniae	Benzylpenicillin (cefotaxime or chloramphenicol if penicillin allergic or penicillin resistant)

Note that rifampicin is given for 48 h or ciprofloxacin (single dose) to patients with meningitis caused by *N. meningitidis* and *H. influenzae* to eliminate nasopharyngeal carriage.

[a] Empirical treatment will be directed by the most likely causative organism and local antimicrobial sensitivity pattern.

Clinical features

Classic features (fever, meningism) are often absent. Presenting signs are subtle and include temperature instability, lethargy, decreased feeding and irritability.

Treatment

Early empirical therapy is imperative. Penicillin plus an aminoglycoside (e.g. gentamicin) or third-generation cephalosporin is commonly used.

Postoperative meningitis

Meningitis is a serious complication of neurosurgical procedures. Gram-negative bacilli (*E. coli*, *Pseudomonas*) are important pathogens. *Staphylococcus epidermidis* is a frequent cause of CSF shunt infections. The wide range of possible pathogens requires broad-spectrum empirical therapy (e.g. ceftazidime). Vancomycin is often required for the treatment of *S. epidermidis* shunt infections.

Tuberculous meningitis

• *Mycobacterium tuberculosis* is a rare cause of bacterial meningitis in developed countries, but remains an important complication of TB in many areas of the world. It is a complication of either primary TB (normally pulmonary) or post-primary reactivation. Presentation is often indolent, with headache, malaise, anorexia, photophobia, neck stiffness and decreasing level of consciousness.

• Microscopy of CSF shows a lymphocytosis, high protein count and low CSF glucose (see Table 38.2). A Ziehl–Neelsen stain of the CSF may show acid-fast bacilli, but definitive diagnosis is often reliant on CSF culture.

• Treatment follows the principles of anti-tuberculous triple therapy (rifampicin, isoniazid and pyrazinamide).

• Long-term complications are frequent because of the dense fibrous exudate; steroids appear to reduce complications.

Viral meningitis

Viral meningitis, the most common form of meningitis, is often mild and full recovery is usual.

Aetiology

Important causes are enteroviruses (echoviruses, Coxsackieviruses, polioviruses, enterovirus 71) mumps in countries without routine childhood immunization coverage (MMR), herpes simplex particularly in association with primary genital herpes (4-8% of cases) and arboviruses in countries where they are endemic.

Clinical features

Influenza-like illness followed by meningism (neck stiffness, headache and photophobia).

Laboratory diagnosis

This is by: CSF microscopy which shows a lymphocytosis (see Table 38.2); viral culture of CSF, stool and throat swabs and detection of viral nucleic acids by molecular techniques (PCR).

Treatment

There is no specific antiviral therapy, except aciclovir for herpes simplex.

Fungal meningitis

Fungal meningitis occurs principally in immunocompromised patients, but can occur rarely in immunocompetent patients. Pathogens include *Cryptococcus neoformans*, *Aspergillus* spp. and *Candida* spp.

Protozoan meningitis

Very rarely, the free-living amoeba, *Naegleria fowleri*, can cause meningitis. Patients normally have a history of swimming in warm, brackish water 1–2 weeks before presentation. Direct invasion of the meninges via the cribriform plate results in severe meningitis with a high mortality. Amoebae may be seen in unstained CSF preparations. Treatment is with amphotericin.

Encephalitis

Encephalitis is mainly caused by viruses, but other organisms are occasionally responsible (Table 38.4). Clinical signs include fever and vomiting, followed by decreased level of consciousness, focal neurological signs and eventually coma. Some important causes are outlined below; other causes are described in chapters dealing with individual pathogens.

Table 38.4 Principal causes of encephalitis and meningoencephalitis.

ENCEPHALITIS AND MENINGOENCEPHALITIS	
Cause	**Comments**
Viruses:	
Herpes simplex virus	Most common cause of viral encephalitis
Mumps virus	Rare complication of mumps parotitis
Varicella-zoster virus	Complication of chickenpox and zoster
Cytomegalovirus	Immunosuppressed patients (e.g. AIDS, transplant recipients)
Human herpesvirus 6	Rare complication of primary infection in young children and after primary infection or reactivation in immunocompromised patients
Polio- and other enteroviruses	Important cause in developing world
Measles virus	Rarely in acute measles; disease develops 6–8 years later, causing subacute sclerosing panencephalitis (SSPE)
Rabies virus	Zoonosis
Arboviruses	Important cause of meningitis and encephalitis in endemic areas
Retroviruses HTLV-I and HIV	Endemic in some areas of the world; must be distinguished from other causes in AIDS patients
Bacteria:	
Treponema pallidum	Tertiary syphilis
Mycoplasma pneumoniae	Rare complication of mycoplasma pneumonia
Borrelia burgdorferi	Part of multisystem infection (Lyme disease)
Protozoa and fungi:	
Toxoplasma gondii	Immunocompromised people
Cryptococcus neoformans	Meningoencephalitis, mainly immunocompromised individuals
Plasmodium falciparum	Cerebral malaria
Trypanosomes	
Toxocara	Larvae in brain; granulomata form

Herpes simplex virus (HSV)

• Encephalitis is a rare complication of HSV infection. The incidence is increased in immuno-compromised patients, neonates and elderly individuals.

• HSV may complicate primary (infancy) infection or follow viral reactivation (adults). Invasion of the temporal lobe is a prominent feature. Clinical diagnosis is often difficult; electroencephalographic (EEG) studies and computed tomography (CT) may be helpful.

Laboratory diagnosis
CSF may contain a few lymphocytes, but culture for HSV is often negative. The determination of serum antibodies is often unhelpful but detection of CSF antibodies is useful. PCR is probably the most useful diagnostic test for HSV encephalitis. Brain biopsy is the definitive diagnostic test, but is difficult to justify.

Treatment and prevention
Treatment is with intravenous aciclovir. Neonatal HSV encephalitis is normally acquired from the maternal genital tract during labour and pregnant women with active genital lesions at term should undergo caesarean section. Immunocompromised patients, particularly transplant recipients, often receive aciclovir prophylaxis to prevent invasive HSV disease.

Brain abscesses

Aetiology
Causative organisms related to pathogenesis; important causes include *Streptococcus milleri, S. aureus* and mixed aerobic/anaerobic infections (e.g. *E. coli* plus *Bacteroides*).

Pathogenesis
May follow trauma, surgery, septic emboli or direct invasion from adjacent infection (e.g. mastoiditis).

Diagnosis
Clinical and radiological; blood cultures or culture of aspirate may define the causative organism(s).

Treatment
Drainage and antibiotics as directed by culture.

Encephalopathy associated with scrapie-like agents

• Degenerative diseases caused by abnormal prion proteins. Prions are proteins found in normal cells but sometimes transform into a pathogenic abnormal form. Infectivity of abnormal prions resists heat and chemicals, including formaldehyde.

• The infectious agent infects a range of mammals, including humans and replicate slowly with a long incubation period (i.e. scrapie in sheep, bovine spongiform encephalopathy [BSE] in cattle).

• Diagnosis is from histological changes in the brain. No treatment or vaccine is available.

Creutzfeldt–Jakob disease (CJD)

This is a rare encephalopathy of humans with an uncertain mode of transmission. Transmission has occurred with corneal grafts, neurosurgery and growth hormone preparations. Variant CJD (vCJD) first reported in the UK is thought to have resulted from transmission of infection from BSE in cattle to humans via contaminated beef. Prions are not destroyed by normal sterilization cycles in autoclaves. Requires 134°C for 18 min or exposure to sodium hydroxide.

Kuru

Transmission is from human to human by cannibalism and recorded mainly in Papua New Guinea.

Tetanus

• Tetanus toxin is produced by *Clostridium tetani* in infected wounds. The toxin is carried to the peripheral nerve axons and CNS, where it blocks inhibition of spinal synapses, resulting in over-activity of motor neurons with exaggerated reflexes, muscle spasms, lock-jaw (trismus), neck stiffness and opisthotonos (spasm of muscle of back, causing arching of trunk).

• Treatment is with anti-tetanus immunoglobulin

and penicillin, and excision of the wound. Respiratory support is needed and mortality is high.

• Prevention is by immunization with toxoid vaccine.

Chronic meningitis

• When the neurological symptoms and signs persist or progress for at least 4 weeks.

• Infections include amoebae, brucellosis, cryptococcal, Lyme disease, syphilis and TB.

• Discrete (focal) lesions can occur in the brain with certain infections including toxoplasmosis and aspergillosis.

• Non-infectious diseases can also give a similar clinical appearance including: neoplastic meningitis, sarcoidosis, connective tissue diseases and chronic lymphocytic meningitis.

Chapter 39

Septicaemia and bacteraemia

Definitions

Bacteraemia

The transient presence of bacteria in the bloodstream; clinical signs of septicaemia are absent, although there may be symptoms and signs of local infection that represent the source of most bacteraemias.

Septicaemia

The presence of bacteria in the bloodstream associated with signs of shock, including tachycardia and reduced blood pressure. Symptoms and signs of localized infection, e.g. pneumonia, osteomyelitis, may also be present.

Endotoxic shock

Lipopolysaccharide (LPS = endotoxin), a component of the cell walls of Gram-negative organisms, can result in endotoxic shock. The presence of endotoxin leads to activation of various inflammatory cascades (Figure 39.1). Release of cytokines results in vasodilatation and permeability, causing a fall in blood pressure (septic shock). Activation of the clotting cascade may result in disseminated intravascular coagulopathy (DIC), with bleeding and thrombosis occurring simultaneously.

Endotoxaemia can occur in the absence of bacteria in the bloodstream, e.g. from Gram-negative bacteria resident in the gut, endotoxin passing across the gut wall. The lipoteichoic acid in the cell walls of Gram-positive organisms, e.g. pneumococci, can activate an inflammatory cascade to produce a clinical picture indistinguishable from 'endotoxic shock'. Greater understanding of the immunological pathways leading to endotoxic shock have led to the design of monoclonal antibodies and drugs that may block the development of endotoxic shock.

Aetiology

Septicaemia is almost invariably a complication of localized infection (e.g. pneumonia, meningitis). The isolation and identification of bacteria from the blood is an important aid to the diagnosis of local infection. The common causes of septicaemia and the likely organisms are shown in Tables 39.1–39.4.

The relative incidence of microorganisms causing septicaemia varies in different countries and with patients' age and circumstances, e.g.:
- Coagulase-negative staphylococci: most frequent blood culture isolate in hospitalized patients in developed countries; related to frequent use of intravascular catheters (Plates 61–63).
- *E. coli*: frequent cause of septicaemia in elderly patients with urinary tract infections.
- *Salmonella* spp.: uncommon in developed countries, but an important cause of septicaemia in developing countries.

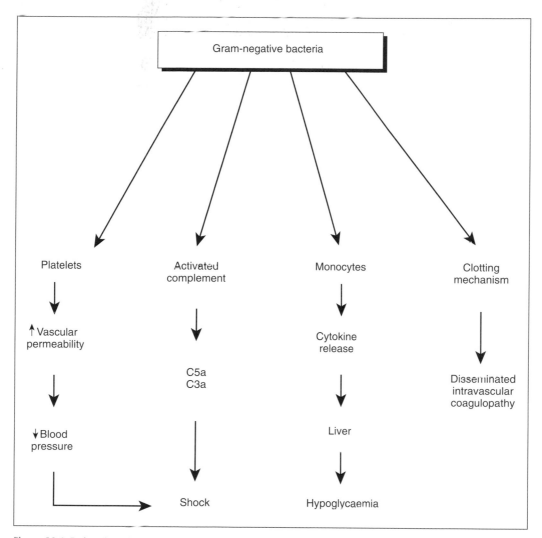

Figure 39.1 Endotoxin activates several immune mechanisms and affects the clotting pathway.

Laboratory diagnosis

• Blood for culture should be obtained aseptically. Bacteria that are part of the normal skin flora (e.g. coagulase-negative staphylococci) may contaminate blood cultures occasionally and must be distinguished from true infection by clinical assessment.

• The traditional method for detecting growth in blood cultures is by regular observation for turbidity. The length of time before growth is detectable depends on the type of microorganism, the number of bacteria and whether antibiotics were present in the original sample; most blood cultures become positive within 48 h. A variety of automated blood culture instruments is now available that detect bacterial growth by various techniques, e.g. the detection of carbon dioxide by radiometric or optical methods.

• When bacterial growth is detected, Gram staining of the blood culture gives a presumptive indication of the likely organism. Further identification and antibiotic susceptibility testing are carried out after the microorganism has been grown.

Table 39.1 Causes of septicaemia: Gram-positive cocci.

GRAM-POSITIVE COCCI		
Organism	Common focus of infection	Notes
Staphylococcus aureus	Deep abscesses, osteomyelitis, septic arthritis, intravascular catheter infection, endocarditis	
Coagulase-negative staphylococci (e.g. Staphylococcus epidermidis)	Intravascular catheter infection; other prosthetic device infections	Frequent contaminant of blood cultures, therefore clinical assessment important in confirming diagnosis
	Endocarditis	Often post-cardiac valve replacement
β-haemolytic streptococcus group A	Cellulitis, necrotizing fasciitis, puerperal sepsis, pharyngitis	Severe acute presentation; high mortality
β-haemolytic streptococcus groups C/G	Cellulitis	
β-haemolytic streptococcus group B	Septicaemia (without focus), meningitis	Important pathogen in neonatal period
Streptococcus milleri	Deep abscesses	Particularly liver, intra-abdominal, lung and brain
Streptococcus pneumoniae	Pneumonia, otitis media, meningitis	More frequent in splenectomized patients
Viridans streptococci	Endocarditis	Often subacute presentation

Table 39.2 Causes of septicaemia: Gram-positive bacilli.

GRAM-POSITIVE BACILLI		
Organism	Common focus of infection	Notes
Diphtheroids	Prosthetic device infection	Less common than Staphylococcus epidermidis
Listeria monocytogenes	Septicaemia (without focus), meningitis	Important in neonates and immunocompromised patients
Clostridium perfringens	Gas-gangrene anaerobic cellulitis	Uncommon, sometimes associated with diabetes mellitus

Table 39.3 Causes of septicaemia: Gram-negative cocci.

GRAM-NEGATIVE COCCI		
Organism	Common focus of infection	Notes
Neisseria meningitidis	Septicaemia (without focus) Meningitis	Often acute and severe with purpuric rash; endotoxic shock a common feature
Neisseria gonorrhoeae	Genital infection	Septicaemia is rare (<3% of cases of gonorrhoea) associated with arthritis and skin rash

Table 39.4 Causes of septicaemia: Gram-negative bacilli.

GRAM-NEGATIVE BACILLI		
Organism	Common focus of infection	Notes
Escherichia coli and other coliforms	Urinary tract infection; intra-abdominal sepsis (e.g. appendicitis, cholangitis); nosocomial pneumonia	
Pseudomonas aeruginosa	Nosocomial infections	Pneumonia in ventilated patients
Haemophilus influenzae type b	Meningitis, epiglottitis	Formerly in children < 5 years; now uncommon with Hib immunization
Salmonella serotypes *Typhi* and *Paratyphi*	Septicaemic illness	Rare in UK, cases normally imported
Other *Salmonella* spp.	Enteritis	Complication of salmonella enteritis; associated with HIV infection
Brucella spp.	Septicaemic illness	Chronic illness often presenting as a pyrexia of unknown origin

Management

Management of septicaemia involves supportive therapy (e.g. intravenous fluids to prevent shock) and antimicrobial agents. The choice of antimicrobial treatment depends on a number of factors.

• The clinical symptoms and signs suggesting the likely focus and cause of infection
• The patient's underlying condition.
• The blood culture results, including the Gram stain (after initial incubation).
• The severity of the patient's condition.
• Patient allergies.
• Severely ill septicaemic patients require high-dose broad-spectrum antibiotics to cover a range of likely pathogens, including those that may be antibiotic resistant.

Antibiotic therapy can be adjusted according to the results of culture, bacterial identification and antibiotic susceptibility. The therapy may also need altering if a patient is in renal or liver failure and depends on how the antimicrobial is metabolized or excreted.

Chapter 40

Device-related infections

Introduction

Modern clinical practice is highly dependent on the use of temporary or permanent medical devices inserted into the patient's body. Medical devices are foreign bodies and are associated with specific infectious complications. Microorganisms may gain access to the device during surgical implantation, after implantation or via haematogenous spread from a distant focus. Common medical devices widely used in clinical practice are shown in Table 40.1.

Pathogenesis of device-related infection

Device-related infection (DRI) results from a combination of bacterial, host and device factors:

- *Bacterial factors*: access to the device by a microorganism is the first stage in development of DRI. This can commonly occur at the time of insertion of the device or via local (e.g. from an associated wound) or haematogenesis (blood-borne) spread. After access, the microorganism needs to adhere to the device. Non-specific forces, e.g. van der Waal's; hydrophobic molecules e.g. glycoproteins and specific bacterial adhesions e.g. pili, fimbriae, may assist bacterial adhesion. Some bacteria, e.g. *Staphylococcus epidermidis*, secrete a polysaccharide 'slime' (front cover) that facilitates adhesion and cell aggregation. Bacteria grow-

ing on devices commonly form a biofilm, which consists of polysaccharide (glycocalyx) produced by the microorganism. The microorganisms commonly associated with DRI are shown in Table 40.2.

- *Host factors*: may increase risk of DRI; include immunosuppression, diabetes, site of prosthesis.
- *Device factors*: these include: *source of device material*, e.g. synthetic, biomaterial; *surface of device*, e.g. textured, smooth, hydrophobic, hydrophilic; *shape of device*, e.g. tubing, mesh.

Intravascular catheter-related infection including central venous catheters (CVCs) and peripheral catheters

Clinical features
Localised and/or systemic infection.

Localised infection
Pain, swelling, erythema (Plates 62 & 63) and purulent discharge at the catheter exit site.

Systemic infection
Fever usually low grade (≤38.5°C), rigors on flushing catheter, no other obvious focus of infection.

The clinical diagnosis of intravascular catheter-related infection may therefore be difficult to establish as a result of the non-specific clinical presentation.

Table 40.1 Examples of some medical devices commonly used in medicine

COMMONLY USED MEDICAL DEVICES	
Medical device	**Example of clinical use**
Central venous catheters	Haemofiltration/haemodialysis
	Administration of drugs, fluids, chemotherapy; venous blood sampling; total parenteral nutrition
	Drug and fluid administration
Peripheral venous catheters	Fluids; administration of drugs
Continuous ambulatory peritoneal dialysis (CAPD) catheters	Renal dialysis
Urinary catheters	Maintenance of urine flow
Prosthetic joints	Replacement joints, e.g. hip, knee
Prosthetic heart valves	Replacement of failing heart valves
CSF shunts	Intracranial pressure control
Epidural catheters	Administration of fluids, e.g. anaesthetic, steroids
Endotracheal tubes	Provision of artificial ventilation

Table 40.2 Examples of microorganisms commonly associated with device-related infections

MICROORGANISMS COMMONLY ASSOCIATED WITH DEVICE-RELATED INFECTIONS	
Device	**Common pathogens**
Catheters (CVC, peripheral, CAPD)	Coagulase negative staphylococci (CNS), *Staphylococcus aureus*
Urinary catheter	Coliforms, *Candida albicans*
Prosthetic heart valves	CNS , *Staphylococcus aureus,* α-haemolytic streptococci
Prosthetic joints	CNS, *Staphylococcus aureus*
CSF shunts	CNS
Epidural catheters	CNS, *Staphylococcus aureus*
Endotracheal tubes	Coliforms including *Escherichia coli* and *Klebsiella pneumoniae.* Also *Pseudomonas aeruginosa* and *Staphylococcus aureus*

Laboratory diagnosis
- *Localised infection*: culture of exit site swab (high negative predictive value).
- *Systemic infection:* semi-quantitative/quantitative culture of explanted CVC distal tip or peripheral catheter. If systemic CVC infection is suspected paired blood cultures (via separate peripheral venepuncture and CVC) should be taken for quantitative/qualitative analysis.

Treatment
Depends on causative microorganism; if *Staphylococcus aureus, Candida albicans*, coliforms, then appropriate antimicrobial and catheter removal; if CNS then attempt antibiotic treatment without catheter removal.

Prevention
Good aseptic technique pre- and post-insertion may aid in reducing intravenous catheter infection.

Continuous ambulatory peritoneal dialysis (CAPD) catheter infections

Clinical features
Localized infection at catheter insertion site; may include: pain, swelling, erythema, sometimes associated with purulent discharge. If associated with peritonitis may present with abdominal pain, pyrexia, nausea, vomiting and tachycardia. Usually have a raised dialysate white cell count (WCC > 100/mm^3)

Laboratory diagnosis

Localized infection; exit site swab culture. Systemic infection; culture peritoneal dialysis fluid (PDF); WCC in PDF usually elevated.

Treatment

Depends on causative microorganism and their antibiotic sensitivity. Examples include: vancomycin (intraperitoneal) if CNS; piperacillin–tazobactam for sensitive coliforms.

Prevention

Strict aseptic technique during CAPD bag exchange including disinfection of the hands and CAPD tubing connections with 70% alcohol.

Urinary catheter infection

Clinical features

Urinary tract infection (see Chapter 36).

Laboratory diagnosis

See Chapter 36.

Treatment

Removal/replacement of catheter and appropriate antimicrobial treatment (see Chapter 36).

Prosthetic joint infection

Clinical features

Joint or bone pain, tenderness, inflammation with or without swelling, fever. Complications may include loosening of the prosthetic joint, sinus formation, low-grade bacteraemia, chronic infection unless the prosthesis is removed.

Clinical diagnosis

To differentiate prosthetic joint infection from mechanical loosening is difficult.

Laboratory diagnosis

Non-specific laboratory tests, elevated differential WCC, raised C-reactive protein (CRP) and erythrocyte sedimentation rate (ESR).

Microscopy

Gram stain of joint material (aspirates) or, if undergoing operative procedures, tissue, e.g. acetabular, capsular, femoral tissue for hip operations.

Culture

Joint fluid, tissue and blood should be cultured. However, as many pathogens are derived from skin flora. At least two consecutive positive cultures from different tissues or aspirates assist in establishing the diagnosis.

Treatment

Usually requires removal of prosthesis and antibiotic treatment. Requires several weeks of antibiotics if infection involves bone. Antibiotics include vancomycin for flucloxacillin-resistant CNS.

Prevention

Appropriate antibiotic prophylaxis for implant surgery. Pre-existing infections (skin, urinary tract, oral cavity) may result in prosthetic joint infection via haematogenous seeding postoperatively and therefore require treatment before surgery. Intraoperative measures to prevent bacterial contamination at the surgical site are paramount (e.g. ultra-clean filtered air; surgical gowns; aseptic techniques, including impervious drapes and double gloving).

Prosthetic heart valve infection

Clinical features

Early (< 60 days after valve implantation) and late (> 60 days) endocarditis. Symptoms include fever, chills, anorexia, weight loss and night sweats.

Clinical diagnosis

Made using Duke's criteria (includes presence of vegetation on heart valve and positive blood cultures). Infection may be associated with the presence of intravascular catheters.

Laboratory diagnosis

Non-specific laboratory tests, ESR, CRP, WCC, proteinuria and microscopic haematuria. Blood cultures. A minimum of at least three sets of blood cultures should be taken before antibiotic therapy. Commonly several sets are culture positive.

Cardiological investigation
Echocardiograph.

Treatment
Depends on causative pathogen.

Prevention
Strict intraoperative measures (as for prosthetic joint infection); prophylaxis for specific procedures.

Epidural catheter-associated infections

Immunosuppression (diabetes, HIV, alcoholism) is a predisposing factor for infection. Epidural infection may cause extensive damage to the associated structures. Spinal epidural abscesses may occur.

Clinical features
Early presentation may be subtle and clinical diagnosis difficult to establish. Symptoms include fever; localized back pain, neurological deficit.

Laboratory diagnosis
Non-specific tests (CRP, ESR, WCC) and blood cultures.

Treatment
Commonly associated with CNS that needs intravenous vancomycin therapy if resistant to flucloxacillin.

Prevention
Good aseptic technique on insertion. Antibiotic treatment to eradicate existing infections, which may lead to haematogenous seeding of the device.

CSF shunts

Clinical features
CSF shunts are commonly inserted via the ventriculoperitoneal (VP) route or through the ventriculoatrial (VA) route. Average incidence of infection is approximately 10%.

VA shunt infections often do not appear for months/years after insertion. Early cases are characterized by fever, tachycardia and rigors. Delayed cases present with intermittent/rare fever, rash arthralgia, anaemia and muscle aches.

VP infection presents within a few months of operation. Symptoms include vomiting, headache and visual disturbance, sometimes accompanied by abdominal distension.

Laboratory diagnosis
Blood culture; cerebrospinal fluid (CSF) for microscopy and culture. Non-specific laboratory tests including CRP and ESR may provide supporting evidence of infection.

Treatment
Shunt removal and treatment with antimicrobials according to causative pathogen.

Prevention
Aseptic techniques on insertion and antimicrobial prophylaxis.

Chapter 41

Drug-resistant microorganisms

Introduction

The use of antimicrobial drugs inevitably leads to the selection and spread of resistance in microorganisms. Microorganisms are able to evolve and disseminate more rapidly than new antimicrobials can be developed. To prevent a return to the 'pre-antibiotic era' requires the judicious use of antimicrobials in humans and animals, and the prevention of the spread of resistant organisms by hygiene measures.

This chapter describes some resistant microorganisms that present clinical problems at present.

Meticillin-resistant *staphylococcus aureus*

Epidemiology
- Meticillin-resistant ***Staphylococcus aureus*** (MRSA) was first described in the 1960s shortly after meticillin was introduced to overcome penicillin-resistant *S. aureus*. Numbers at this time were very small.
- MRSA re-emerged in the 1980s primarily in hospitals and multiple antibiotic-resistant *S. aureus*. MRSA is now endemic in most British hospitals.
- There are a number of clones of MRSA; those most prevalent ('epidemic' MRSA) in the UK are EMRSA 15 and EMRSA 16. Although these clones of MRSA are considered a hospital-acquired organism, they may be acquired and transmitted in the

community. These MRSA clones have the same range of pathogenicity as meticillin-sensitive *S. aureus* (MSSA).
- Most recently, 'community-acquired MRSA' (CA-MRSA) has now been described in patients with no links to hospitals. Unlike the 'hospital-associated' MRSA, CA-MRSA tends to be sensitive to antibiotics other than β-lactams. CA-MRSA is, however, more likely to cause serious infections as a result of the presence of Panton–Valentine leukocidin.
- Some countries, notably those in Scandinavia, have an extremely low prevalence of MRSA.

Mechanism of resistance and detection
- Resistance in MRSA is caused by the *mecA* gene, which encodes a low-affinity penicillin-binding protein. This results in resistance to all current penicillins, cephalosporins and carbapenems. The gene *mecA* may have been acquired from coagulase-negative staphylococci.
- Detection is by conventional antibiotic susceptibility tests. The media used must allow expression of the *mecA* gene, so that meticillin resistance is phenotypically manifest. Detection is also possible by polymerase chain reaction (PCR) for *mecA*.

Control
- Hand hygiene: the most important; prevents up to 50% of hospital-acquired infections; decreases the chance of organisms being carried from patient to patient via the hands of healthcare workers.

- Environmental cleanliness: reduces the likelihood of patient acquiring organisms directly from the environment, or the hands of healthcare workers becoming contaminated from the environment.
- Equipment decontamination: as per environmental cleanliness.
- Screening of *S. aureus* carriage sites to identify carriers who may be asymptomatic.
- Patient isolation/cohorts: reinforces the need for hand hygiene; physical containment of MRSA in the environment.
- Patient decontamination: antiseptics applied to the nose (mupirocin) and skin (e.g. chlorhexidine body washes) to reduce carriage and contamination of healthcare workers' hands; may reduce shedding of organisms into the environment.
- Prophylactic antibiotics to prevent infection (from MRSA or MSSA) in certain forms of surgery, e.g. implants.
- The Department of Health (in England) uses MRSA bacteraemia rates as a measure of infection control performance, and publishes league tables of these rates.

Treatment
- Glycopeptides (vancomycin, teicoplanin) are the mainstay of serious MRSA infections. True vancomycin resistance in MRSA is extremely rare. The few cases that have been described in the USA were caused by the acquisition of the *vanA* gene from enterococci (see below).
- Depending on the clone and therefore the repertoire of resistance in MRSA, macrolides (erythromycin), tetracyclines or trimethoprim may be used as oral agents.

Vancomycin-insensitive *S. aureus*

- Vancomycin-insensitive *S. aureus* (VISA or GISA—glycopeptide-insensitive *S. aureus*) was first recognized in the late 1990s in patients who had received multiple and prolonged courses of vancomycin. The level of vancomycin resistance is not sufficiently high to be truly resistant, and for this reason it is termed 'insensitive'; however, it may result in clinical failures of treatment with vancomycin.

- Insensitivity to vancomycin is the result of the effect of an abnormally thick cell wall. It may be difficult to detect by conventional methods. For this reason, the prevalence of VISA is not known. Detection of VISA is by time-consuming killing experiments using vancomycin at different concentrations.
- Treatment is use of antibiotics other than vancomycin, guided by sensitivity tests.

Vancomycin-resistant enterococci

Epidemiology
- Vancomycin-resistant enterococci (VREs or GREs—glycopeptide-resistant enterococci) emerged in the 1990s. This may have been the result of the use of glycopeptides in animal husbandry or the use of oral vancomycin to treat the increasing prevalence of *Clostridium difficile* disease. The latter may have been caused by greater use of broad-spectrum cephalosporins.
- VRE are mostly recognized in specialist units with high use of vancomycin, e.g. renal, haematology and oncology units.

Mechanism of resistance and detection
- VREs are classified as VanA, VanB or VanC phenotypes; these result from alterations in the cell which reduce the interaction of glycopeptides with their target. *Enterococcus faecalis* and *E. faecium* with VanA have high-level inducible resistance to vancomycin and teicoplanin. VanB strains have moderate- to high-level resistant to vancomycin and are susceptible to teicoplanin. VanC results in low-level vancomycin resistance in *E. gallinarum* and *E. casselifavus*.
- With care, glycopeptide resistance may be detected by conventional susceptibility tests. PCR can detect the resistance genes.

Control
Hand hygiene, equipment and environmental decontamination and patient isolation (see above).

Treatment
Amoxicillin can be used for VREs that are sensitive to it, and teicoplanin for VanB or VanC phenotypes

and linezolid or Synercid (quinupristin with dalfo-pristin) for VanA.

Extended spectrum beta-lactamases (ESBLs)
The term ESBL is used to describe organisms that have a β-lactamase which confers resistance to a range of β-lactam antibiotics. The ESBL producers destroy the cephalosprins and α-penicillins. In addition, many are multiresistant to non-β-lactam antibiotics including the quinones (e.g. ciprofloxacin) narrowing further the treatment options. Some ESBL producing organisms, e.g. *Acinetobacter* spp., can cause outbreaks with a high related mortality. The predominant organisms which produce ESBL include *Acinetobacter* spp., *Escheria coli* and *Klebsiella* spp.

Tuberculosis (TB)

Epidemiology
• Selection of drug resistance by chromosomal mutation can occur as a result of: inconsistent or partial treatment (when patients start to feel better and do not take all their drugs regularly for the required period); incorrect prescription of combination therapy; or unreliable drugs or drug supply. For these reasons, resistance is most common in the developing world. However, resistance is being found in Europe, particularly related to HIV.
• Resistance to isoniazid and rifampicin are the most clinically significant.

Detection of resistance
Resistance can be detected by standard mycobacterial susceptibility testing methods. In addition, isoniazid, ethambutol and pyrazinamide resistance may be detected more rapidly by PCR.

Prevention and control
• Early detection (sputum Ziehl–Neelsen [ZN] staining), especially among those patients with HIV.
• Standardized short-course chemotherapy under proper case management conditions, including direct observation of treatment (DOTS).
• Human and financial resources to make TB control part of a national health programme, including a reporting system to monitor individual patient and programme outcomes.
• An uninterrupted supply of high-quality drugs.

Treatment
• Treatment of multi-drug-resistant TB may require antibiotics for 18–24 months.
• Antibiotic choice will depend on the sensitivities of the isolates of TB and may include fluoroquinolones (e.g. ciprofloxacin or moxifloxacin), aminoglycosides (streptomycin or amikacin) and capreomycin.
• Treatment is best initiated and followed up at specialist centres.

Candida species

Epidemiology
Resistance to antifungals in *Candida* spp. is increasingly recognized as important in specialist units, particularly those that care for immunocompromised patients. It arises in such units in two ways: intrinsic resistance (species dependent) and acquired during treatment. Cross-infection with resistant *Candida* spp. may also occur.

Detection and treatment
• Susceptibility testing for fungi is difficult and usually carried out at specialist laboratories. Results may therefore be delayed. Hence, differentiation into *C. albicans* (germ tube positive) and other *Candida* spp. is important in deciding which antifungal should be used immediately.
• *C. albicans* isolates are usually sensitive to azoles, amphotericin B and echinocandins (e.g. caspofungin). However, resistance to azoles may occur during or after previous azole treatment. This is a result of the increased production by yeast cells of a drug efflux pump. If patients have recently received or are receiving an azole, and are diagnosed with invasive candidiasis, it is prudent to use another class of drug (amphotericin B or echinocandin) until sensitivity test results are available.
• Species other than *C. albicans* have variable susceptibility to azoles. *C. glabrata* and *C. krusei* are frequently resistant to azoles and sometimes have reduced sensitivity to amphotericin B. *C. lusitaniae*

may be resistant to amphotericin. These species are usually sensitive to caspofungin.

Resistant human immunodeficiency virus

Epidemiology and mechanism of resistance
- Primary reasons for treatment failure include poor drug compliance, pharmacological factors and drug resistance.
- Resistance to drugs for treating HIV has developed against anti-retroviral drugs of all classes (see Chapter 31).
- Drug resistance develops as a result of single or multiple mutations in the viral genome. Some mutations will confer cross-resistance to other drugs in the same class.
- Such mutations exist within the viral population even before treatment starts. The high rate of HIV replication, and the inherent error rate of the viral polymerase enzyme, result in the generation of several viral genetic variants (quasi-species).
- The use of antiviral drugs will provide a selective pressure for the preferential growth of resistance variants that will dominate the viral population.
- The emergent drug-resistant viruses will be the fittest in the presence of drugs but generally they replicate less well than the wild-type virus in the absence of drugs.

- Transmission of drug-resistant viruses has been demonstrated in up to 20% of newly diagnosed cases in Europe and North America.

Detection of resistance
- Resistance to anti-retrovirals can be assessed phenotypically and/or genotypically.
- Phenotypic assays depend on growth in cell culture of isolates in question in the presence of drugs. However, these assays are very slow, expensive and labour intensive, and generally not used in routine clinical practice.
- Genotypic assays rely on the detection of genetic mutation within the target genome and they are the most commonly used methods for HIV drug resistance.
- Several variations of genotypic assays are available, including the use of specific probes to detect individual mutations and nucleic acid sequencing.

Prevention and control
- Compliance with treatment regimens.
- Monitoring of antiviral treatment with regular viral load measurements.
- Change of anti-retrovirals should be guided by the results of resistance assays, if treatment failure is recognized.
- Prevention of HIV transmission (see Chapter 47).

Chapter 42

Cardiovascular infections

Infective endocarditis

Definition
Infection of the endocardium of the heart, usually involving the valves.

Epidemiology
Infective endocarditis is uncommon (approximately 2000 cases per year in the UK). The mortality rate is around 20% but is increased in cases diagnosed late. In about two-thirds of patients there is a pre-existing heart defect, either congenital or acquired.

Pathogenesis
Important causes of valvular damage include rheumatic fever and congenital abnormalities; a damaged or prosthetic valve encourages deposition of fibrin, to which microorganisms can attach during episodes of transient bacteraemia, e.g. after operative procedures such as dental manipulation or gastrointestinal surgery. This forms, together with platelets and fibrin, an irregular mass on a valve called a vegetation.

Clinical features
• Infective endocarditis may be an acute rapidly progressive disease, or subacute with a slow indolent course; it is characterized by fever, night sweats, malaise, anaemia and anorexia.
• Circulating immune complexes result in skin lesions: splinter haemorrhages under the nails and tender nodular lesions on the palms or fingers (Osler's nodes) and glomerulonephritis (microscopic haematuria).
• Infective emboli can result in infarction of various organs, including kidneys, intestines or brain. Rarely, these may be the presenting feature. Septic emboli can also cause macular lesions on the palms and soles (Janeway's lesions), which consist of bacteria, necrosis and haemorrhage. Damage to the valve may result in a changing murmur, cardiac failure and valve perforation.
• The two major criteria to support the diagnosis is, the presence of a vegetation on a valve and positive blood cultures with an organism known to cause endocarditis.

Causative organisms
Important causes of infective endocarditis and their sources are shown in Table 42.1. Other microorganisms, including *Escherichia coli*, other coliforms and *Pseudomonas* spp. are rare causes of infective endocarditis. 'Culture-negative endocarditis' is used to describe cases of infective endocarditis diagnosed clinically in which no causative organism is cultured. This may be the result of antibiotic treatment before investigation or of organisms that do not grow readily in routine laboratory culture media, e.g. *Coxiella burnetii* (the cause of Q-fever) and *Chlamydia psittaci* (the cause of psittacosis).

Table 42.1 Major microbial causes of infective endocarditis

MAJOR MICROBIAL CAUSES		
Organism	**Source**	**Notes**
Enterococci	Gastrointestinal tract	
Staphylococcus aureus	Skin; local infection	Associated with intravenous drug abuse; may occur on a normal valve
Staphylococcus epidermidis	Skin	Leading cause of prosthetic valve endocarditis, particularly in first few months after surgery
Viridans streptococci, e.g *Streptococcus sanguis, S. mitis*	Oral cavity	
Candida albicans	Oral cavity; skin	Has been associated with drug abuse

Table 42.2 Antimicrobial treatment of infective endocarditis.

EXAMPLES OF ANTIMICROBIAL TREATMENT	
Organism	**Treatment**
Penicillin-sensitive streptococci	Benzylpenicillin
Reduced penicillin-sensitive streptococci and enterococci	Benzylpenicillin plus gentamicin
Flucloxacillin-sensitive staphylococci	Flucloxacillin
Flucloxacillin-resistant staphylococci	Vancomycin plus rifampicin
Fungi	Amphotericin
Chlamydia psittaci/Coxiella burnetii	Tetracycline

Laboratory diagnosis

• *Non-specific tests*: a raised white cell count (WCC) and increased inflammatory markers (erythrocyte sedimentation rate [ESR] and C-reactive protein [CRP]) support the diagnosis of endocarditis.

• *Cardiological investigations*: echocardiography, either transthoracic or via an electrode placed down the oesophagus (transoesophageal echocardiography) allows the visualization of vegetations on the cardiac valves.

• *Microbiological investigations*.

— blood cultures: the number of bacteria circulating in the blood is often small in endocarditis and at least three separate blood culture sets should be taken before antibiotic therapy is commenced

— serological tests. *Chlamydia psittaci* and *Coxiella burnetii* are diagnosed by serological tests.

Management (Table 42.2)

The principles of antibiotic therapy in endocarditis are:

• to use combinations of bactericidal antibiotics, which preferably show synergy for prolonged periods (4–6 weeks)

• to monitor therapy carefully by following clinical response and levels of inflammatory markers (ESR, CRP)

• to perform extended microbiological tests on the causative organism including minimum inhibitory concentrations.

Management of cardiac aspects of the condition may involve treatment for cardiac failure and, in some cases, surgery to replace the infected, damaged valve.

Prevention

Prophylactic antibiotics are used for high-risk groups (e.g. previous rheumatic fever, prosthetic heart valve, congenital heart disease) undergoing certain operative procedures, including dental operations, which have a high risk of bacteraemia.

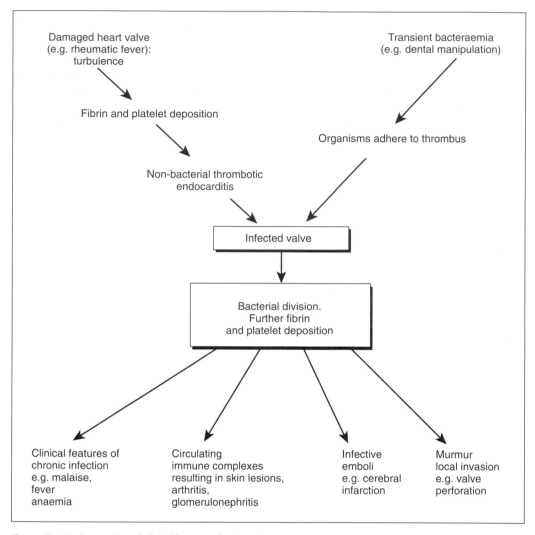

Figure 42.1 Pathogenesis and clinical features of endocarditis.

Myocarditis

Definition
Infection of the myocardium.

Aetiology
Coxsackie B viruses are the most common causes of viral myocarditis; others include cytomegalovirus (CMV), Epstein–Barr virus (EBV) and mumps virus.

Direct bacterial infection rarely causes myocarditis. Diphtheria toxin and toxoplasmosis are also rare causes of myocarditis.

Clinical features
Fever, chest pain, dyspnoea; chest radiograph may show an enlarged heart; a pericardial effusion is often present; electrocardiographic (ECG) changes are a frequent feature.

Laboratory diagnosis

This is by:

• viral isolation from pharyngeal washings or faeces

• myocardial biopsy which may show typical histological features and occasionally viruses that may be grown from the tissue

• serology (a fourfold rise in antibody titres or evidence of an IgM response).

Treatment

Antiviral therapy is available for some causes, e.g. ganciclovir for CMV infection.

Pericarditis

Definition

Inflammation of the pericardium resulting from infection.

Aetiology

• *Viral*: enteroviruses (Coxsackieviruses A and B, echoviruses), CMV, EBV; occasionally, other viruses.

• *Bacteria*: S. aureus, S. pneumoniae, β-haemolytic streptococci, *Mycobacterium tuberculosis*. Bacterial pericarditis may be the result of haematogenous spread from a distant primary site of infection or, less commonly, direct extension from a primary lung infection or after cardiac surgery.

• *Fungal*: *Histoplasma*; *Aspergillus* and *Candida*

spp., as part of invasive fungal disease in immuno-compromised individuals.

Clinical features

These include fever, chest pain, dyspnoea; a pericardial rub (before the pericardial effusion develops); and signs of cardiac failure. Chest radiograph may show an enlarged heart with a pericardial effusion. Patients with bacterial pericarditis normally have an acute presentation and are generally more toxic than patients with viral pericarditis.

Laboratory diagnosis

• *Viral*: isolation from pharyngeal washings, faeces or pericardial fluid; serology.

• *Bacterial*: culture of pericardial fluid and blood cultures.

Management

• *Viral*: specific antiviral therapy, e.g. ganciclovir for CMV disease; pericardial drainage may be necessary.

• *Bacterial*: pericardial drainage and occasionally pericardectomy together with appropriate antibiotics, depending on pathogen.

• *Tuberculous*: drainage and pericardectomy; anti-tuberculous drugs plus corticosteroids to reduce inflammation.

• *Fungal*: amphotericin; pericardial drainage may be necessary.

Chapter 43

Bone and joint infections

The aetiology, diagnosis and management of osteomyelitis (infection of bone) and septic arthritis are similar. Both conditions can present as acute or chronic infections.

Osteomyelitis

Introduction

Infections of bone; can involve all areas of bone and, in children, mainly the growing ends of long bones. Infection can occur via:
- haematogenous spread
- directly from surrounding infected tissue
- direct inoculation after trauma (e.g. open fracture)
- osteomyelitis secondary to spread from an adjacent area of infection is more common in older patients (includes wound infections after orthopaedic surgery, mastoiditis), patients with peripheral vascular disease or diabetes
- often subacute.

Acute osteomyelitis

Aetiology
The microbial aetiology of acute osteomyelitis is shown in Table 43.1.

Clinical presentation
There is high fever with raised white cell count (WCC), erythrocyte sedimentation rate (ESR) and swelling, pain and tenderness over affected bone; overlying cellulitis is common. In infants, reduced use of the affected limb is an important sign. Radiological changes are often not apparent until 1–2 weeks after clinical presentation.

Laboratory diagnosis
Blood cultures are positive in about 40% of cases. Direct aspiration of bone is performed occasionally. Intraoperative specimens may be taken.

Treatment
Antibiotic treatment is normally for at least 6 weeks. Surgery to allow drainage and débridement of dead bone may be necessary.

Chronic osteomyelitis

Aetiology (Table 43.2)
Often results from spread from a local adjacent infection, e.g. chronic otitis media, mastoiditis or periodontal sepsis.

Clinical manifestations
Often these are related to the precipitating infection. Bone involvement may only become apparent following radiological changes. Magnetic resonance imaging (MRI), computed tomography (CT) or technetium bone scans may help identify an infected area but may not exclude other pathologies.

Table 43.1 Causative organisms of acute osteomyelitis.

ACUTE OSTEOMYELITIS		
Organism	Treatment	Comment
Staphylococcus aureus	Flucloxacillin and fusidic acid	Most common cause
Streptococcus pyogenes	Benzylpenicillin	Infrequent cause
Streptococcus pneumoniae	Benzylpenicillin	Usually direct spread from ear or sinus infection
Clostridium perfringens	Benzylpenicillin	Usually post-traumatic
Haemophilus influenzae type b	Cefotaxime	Rare due to Hib vaccine
Salmonella spp.	Ciprofloxacin	Rare cause in developed countries

Table 43.2 Causative organisms of chronic osteomyelitis.

Organism	Comment
Staphylococcus aureus	Most common cause
Streptococcus milleri	Associated with abscesses
Streptococcus pneumoniae	Ear or sinus infection
Actinomyces israelii	Peridontal sepsis
Pseudomonas spp.; coliforms	Contamination of open wounds
Salmonella spp.	Associated with sickle cell disease and septicaemia
Mycobacterium tuberculosis	Rare but important cause
Bacteroides spp.	Associated with chronic otitis media and mastoiditis

Laboratory diagnosis

As for acute osteomyelitis.

Treatment

Antibiotic treatment depends on the causative organisms. Surgical débridement of dead tissue is necessary more frequently than with acute osteomyelitis.

Tuberculous osteomyelitis

• Tuberculous osteomyelitis results after haematogenous dissemination (bloodstream spread) of the organism during the early phases of infection.

• Presentation is insidious with no acute signs; pain is the most common symptom. A discharging sinus may be present occasionally. The most frequent sites are the weight-bearing bones, particularly the spine.

• Investigations include microscopy (Ziehl–

Neelsen stain) and culture of fine-needle biopsy or surgical specimens.

• Treatment with anti-tuberculous drugs is normally sufficient, although, occasionally, surgical intervention is required.

Osteomyelitis after trauma and surgery

Severe compound fractures may become infected with skin bacteria (e.g. staphylococci) or environmental organisms (e.g. clostridia). Contaminated compound fractures may require débridement of devitalized tissue and antibiotic therapy.

Osteomyelitis after orthopaedic operations is often caused by *S. aureus*, but, when associated with prosthetic joint replacements, is more frequently caused by coagulase-negative staphylococci (CNS), e.g. *S. epidermidis* with a characteristic insidious onset and minimal symptoms. Antibiotic treatment is difficult because many CNS are often resistant to flucloxacillin. Vancomcyin is used in this situation. Usually, an infected prosthetic joint needs to be removed and replaced. This may be a two-stage procedure: infected joint removal followed by antibiotics for 4-6 weeks, followed by joint replacement. Prevention of orthopaedic prosthetic device infection involves the use of specially designed operating theatres ventilated with ultra-clean air and perioperative prophylactic antimicrobials.

Septic arthritis

Infection of joints may arise from haematogenous spread or after direct spread from an overlying skin lesion or area of osteomyelitis.

Aetiology
• *S. aureus*: 90% of cases.
• *Haemophilus influenzae* type b: a common cause in children aged < 5 years until the introduction of the Hib vaccine.
• Less common causes include β-haemolytic streptococci, salmonellae, *Neisseria gonorrhoeae*, *Streptococcus pneumoniae* and *Mycobacterium tuberculosis*.

Clinical presentation
There is a hot, swollen joint with fever (Plate 64).

Tuberculous arthritis presents insidiously with a cold swollen joint.

Laboratory diagnosis
Blood cultures are positive in about 50% of cases. Joint aspiration is frequently performed for microscopy and culture.

Management
Antibiotic therapy is required (see 'Osteomyelitis') and surgical drainage may be necessary.

Reactive arthritis

This is immunologically mediated, resulting from infection at a distant site, e.g. *Yersinia enterocolitica* gastroenteritis or non-specific urethritis caused by *Chlamydia trachomatis*. Usually, more than one joint is affected.

Viral arthritis

A number of viruses may cause arthritis as part of a generalized infection, e.g. hepatitis B virus, HIV, rubella virus, mumps virus and parvovirus.

Chapter 44

Skin and soft-tissue infections

Normal flora of skin

The normal skin is colonized mainly by Gram-positive organisms (e.g. *Staphylococcus epidermidis* and diphtheroids), which can survive the relatively harsh conditions—dryness, acidity and the presence of fatty acids and salt. The normal flora is protective, because other species must compete to survive in this ecological niche.

Pathogenesis of skin and soft-tissue infections

- Superficial skin infections affect the epidermis and dermis (e.g. impetigo, erysipelas).
- Deeper infections affect the subcutaneous tissue (e.g. cellulitis (Plates 65 & 66), fasciitis).
- Infections of skin structures may occur (e.g. hair follicles, folliculitis (Plate 57)).
- Skin infections may arise from pathogens that gain entry through damaged areas of skin or from haematogenous spread.
- Bacterial toxins produced by infections at other sites may also result in skin changes (e.g. group A streptococci causing scarlet fever).

Skin and soft-tissue infections

Folliculitis, furuncles and carbuncles

Aetiology
These are commonly caused by *Staphylococcus*
aureus; perianal lesions are often caused by a mixture of faecal organisms (e.g. *Escherichia coli*, anaerobes, *Streptococcus milleri*).

Pathogenesis
Folliculitis results from infection of a hair follicle. The lesion may expand and develop into a furuncle (boil). Boils may eventually rupture, with drainage of pus, or develop into carbuncles (large loculated abscesses), particularly in people with diabetes. Perianal abscesses are a particular problem in neutropenic patients.

Complications
- Local spread to bones, joints and other deep structures.
- Haematogenous spread, resulting in abscesses in various organs.

Laboratory diagnosis
Culture of aspirated pus.

Treatment
Drainage and antibiotics (e.g. flucloxacillin for *S. aureus*; metronidazole for anaerobes).

Impetigo

Impetigo is a superficial infection of the skin, involving the epidermis.

Aetiology
Group A β-haemolytic streptococcus frequently in association with *S. aureus*.

Epidemiology
Principally associated with childhood, poor socioeconomic conditions and overcrowding.

Clinical features
Yellow crusted lesions develop, mainly on the face around the nose. The lesions spread by autoinoculation.

Laboratory diagnosis
Culture of swabs of lesions.

Treatment
Flucloxacillin is the drug of choice; erythromycin is an alternative.

Acne

Acne is associated with *Propionibacterium acnes*. Lesions occur in sebaceous follicles with plugs of keratin blocking the sebaceous canal, resulting in 'blackheads'. Treatment is by tetracycline combined with topical antiseptics.

Erysipelas

Erysipelas is a streptococcal infection involving the superficial layers of the skin (epidermis and dermis).

Epidemiology
Erysipelas principally occurs in elderly patients.

Pathogenesis
Group A β-haemolytic streptococci gain entry through breaks in the skin.

Clinical features
Erysipelas most commonly affects the face, particularly the cheeks and periorbital areas. The area of erythema shows a clear line of demarcation, which helps to distinguish erysipelas from cellulitis. Fever is common.

Laboratory diagnosis
Isolation of group A streptococci from superficial swabs is often unsuccessful; blood cultures and skin aspirates may be positive in about a quarter of patients. Serological evidence (e.g. anti-streptolysin O [ASO] titre) of streptococcal infection may be helpful.

Treatment
Benzylpenicillin or erythromycin.

Cellulitis

Cellulitis is an infection of the dermis and subcutaneous fat.

Aetiology
• Group A β-haemolytic streptococci are the most common cause of cellulitis (Plate 65). Rare complications include septicaemia and acute glomerulonephritis. Other β-haemolytic streptococci may also cause cellulitis.
• *S. aureus* may produce superficial cellulitis that is less extensive than that from group A β-haemolytic streptococci. It is usually associated with wound (Plate 66) or abscess formation.
• *Haemophilus influenzae* type b is a cause in children aged < 5 years.
• *Erysipelothrix rhusiopathiae* is an unusual cause of cellulitis, found in meat and fish handlers.
• Non-cholera vibrios (e.g. *Vibrio vulnificus*) are a rare cause of cellulitis complicating wounds that have become contaminated with seawater.
• Anaerobic cellulitis is a complication of bites (oral flora anaerobes) or infection of devitalized tissue (clostridia) after trauma, diabetes mellitus or vascular insufficiency (Plate 67).

Pathogenesis
Bacteria gain entry through a break in the skin, although this is usually not evident at the time of presentation.

Clinical features
Erythematous, hot, swollen skin with an irregular edge is seen (contrast with erysipelas). There is fever and enlargement of regional lymph nodes.

Anaerobic cellulitis may be foul smelling, with skin crepitus and the presence of necrotic skin.

Laboratory diagnosis
Culture of skin swabs/aspirates (positive in about a quarter of cases) and by blood cultures (occasionally positive).

Treatment
• For streptococci/*Erysipelothrix* spp., penicillin is given. Staphylococcal cellulitis requires treatment with flucloxacillin. For this reason, high-dose flucloxacillin (active against streptococci) may be preferred. Clindamycin may be used in penicillin allergy.
• Anaerobes are treated with metronidazole and débridement of devitalized tissue.
• Treatment of *H. influenzae* type b is with cefotaxime.

Synergistic bacterial gangrene

This is an infection of the superficial fasciae and subcutaneous fat, not involving the muscles. It may involve the abdominal wall or male genitalia (Fournier's gangrene).

Aetiology/pathogenesis
Frequently mixed infection occurs involving *S. aureus*, microaerophilic or anaerobic streptococci and, occasionally, Gram-negative anaerobes (e.g. *Bacteroides* spp.). Predisposing conditions include local trauma, abdominal and genital surgery, and diabetes mellitus.

Clinical features
Rapid, spreading cellulitis is present with black necrotic areas. The condition is painful, with a high fever.

Laboratory diagnosis
Culture of affected tissue or pus.

Treatment
Radical excision of the necrotic area is required plus antibiotic therapy: flucloxacillin (*S. aureus*), penicillin (streptococci), metronidazole (anaer-

obes). Broad-spectrum Gram-negative cover, e.g. gentamicin or ciprofloxacin, is also needed.

Necrotizing fasciitis

An infection of the skin and deeper fascia; may involve the muscles.

Aetiology
Group A β-haemolytic streptococci alone. More commonly these infections are usually caused by mixtures of non-group A aerobic streptococci, aerobic Gram-negative bacilli, anaerobic Gram-positive cocci, and *Bacteroides* spp. Onset may follow trauma or recent surgery.

Pathogenesis
Group A streptococci produce enzymes that facilitate tissue spread. Patients with diabetes have an increased risk.

Clinical features
Rapidly spreading cellulitis with necrosis; the patient is toxic with an elevated temperature; high mortality.

Laboratory diagnosis
Culture of affected tissue or pus.

Treatment
Urgent resection of affected tissue plus high-dose benzylpenicillin with metronidazole; clindamycin is an alternative in penicillin-allergic patients. Broad-spectrum Gram-negative cover, e.g. gentamicin or ciprofloxacin, is also needed.

Clostridial cellulitis and clostridial myonecrosis (gas gangrene)

Epidemiology
Now rare in developed countries; peripheral vascular disease and diabetes mellitus are important risk factors.

Pathogenesis
Wounds caused by trauma or surgery can become infected with clostridia, in particular *Clostridium*

perfringens, resulting in gas gangrene or clostridial myonecrosis. Clostridia produce enzymes and exotoxins that facilitate rapid spread through tissue plains and result in haemolysis and shock; gas production results in tissue crepitus.

Clinical features
Cellulitis with necrotic areas and gangrene (Plate 67), later characterized by tissue crepitus, fever and toxaemia as myositis develops.

Laboratory diagnosis
Isolation of clostridia from tissue/swabs or blood. (Note that the isolation of clostridial species from wounds does not define a case of gas gangrene; clostridia can contaminate wounds without infection; gas gangrene is primarily a clinical diagnosis.)

Treatment
Early intervention on suspicion is vital. Penicillin or metronidazole plus débridement of devitalized tissue.

Prevention
Wounds with devitalized tissue should be débrided, cleaned thoroughly and carefully monitored during healing. Prophylactic antibiotics (penicillin) should be prescribed for patients undergoing amputations for peripheral vascular disease.

Venous ulcers/bedsores

A variety of conditions (e.g. diabetes mellitus, peripheral vascular disease, venous insufficiency) may result in areas of skin that have a reduced blood supply and become necrotic, forming ulcers. Ulcers may occur around the foot and ankle (venous ulcers) or the sacral area in bedridden patients (bedsores), and are prone to bacterial colonization and, occasionally, infection.

Typical culture results include:
• mixed growth of coliforms and anaerobes; normally not significant
• *P. aeruginosa*; may reflect antibiotic replacement flora; occasionally requires antibiotic therapy
• *S. aureus*; can be colonization or infection
• β-haemolytic streptococci groups C and G may

colonize venous ulcers and occasionally cause infection; group A β-haemolytic streptococci frequently cause infection.

Distinguishing colonization from infection is important, as antibiotic treatment of colonized ulcers is of no benefit; spreading cellulitis around the ulcer suggests infection. Deep infection including osteomyelitis may occur.

Burns

As with venous ulcers, burns may become colonized with bacteria, e.g. *S. aureus*, *Pseudomonas* spp. and coliforms. The need for antibiotic treatment is dependent on the patient's condition, e.g. the viability of skin grafts, and the organism isolated (group A β-haemolytic streptococci always require treatment). Colonizing organisms may cause bacteraemias.

Wound infections

Aetiology
• Accidental wounds: *S. aureus*, β-haemolytic streptococci, clostridia.
• Bites: oral commensal bacteria, including anaerobes and streptococci; *Pasteurella multocida* (animal bites).
• Surgical wounds (see below).

Laboratory diagnosis
Culture of pus or wound swab.

Treatment
This is dependent on culture results. Prompt prophylactic treatment of human and animal bites with co-amoxiclav is important because organisms may be inoculated into deep tissues, with rapid spread of infection through tissue.

Surgical wound infections

Predisposing factors
• *Underlying conditions*, e.g. diabetes mellitus and peripheral vascular disease, which result in poor blood supply.
• *Type of operation*: surgery involving contaminat-

ed areas (e.g. gastrointestinal tract) or infected tissue (e.g. appendicitis) is more likely to result in postoperative wound infections.
- *Surgical technique*, e.g. poor technique in wound closure, may result in haematoma formation or devitalized tissue, predisposing to wound infection.
- *Postoperative factors*: cross-infection from other infected cases.

Aetiology
- *Upper body*: S. *aureus*, group A β-haemolytic streptococci.
- *Lower body* (particularly abdominal surgery): coliforms, occasionally mixed with anaerobes; *P. aeruginosa*.

Clinical features
Fever; wound exudate; cellulitis.

Laboratory diagnosis
Culture of wound swab or pus.

Treatment
Dependent on causative organism.

Prevention
- *Preoperative preparation*: treatment of sepsis before operation when possible; decontamination of operative area (e.g. bowel preparation) and the use of *prophylactic antibiotics* for operations involving contaminated or infected areas.
- *Good surgical technique* to prevent haematoma formation and wound necrosis and *aseptic theatre practice* with sterile instruments and a clean environment.
- *Measures to prevent postoperative contamination of wounds*, e.g. 'no-touch' dressing technique.

Toxin-mediated skin conditions

Scarlet fever

This is a group A streptococcal throat infection caused by a strain producing an erythrogenic toxin. The toxin causes an erythematous rash with a red ('strawberry') tongue and circumoral pallor. As the rash fades there is desquamation of the skin. Treatment is penicillin or erythromycin.

Staphylococcal scalded skin syndrome

This occurs in small infants (Ritter's diseases) and occasionally in older children and adults (toxic epidermal necrolysis or Lyell's disease). The *S. aureus* strain causes an infection at a distant site but produces a toxin that results in a severe rash with skin desquamation. Diagnosis is on clinical grounds and is supported by the isolation of a toxin-producing strain of *S. aureus*. Treatment is with flucloxacillin. A similar toxin-mediated rash may be seen in *toxic shock syndrome*, also a complication of *S. aureus* infection.

Dermatophyte fungal infections

These have a worldwide distribution, with transmission by direct contact or via fomites (e.g. towels, hairbrushes). Infection may be acquired from animals. See Table 44.1 for aetiology and clinical features.

Laboratory diagnosis
Microscopy and culture of appropriate specimens (skin scrapings, nail clippings, hair roots).

Treatment
This is with topical azole antifungals. When there is nail involvement or, in severe infections when topical azoles have failed, oral terbinafine can be used. Treatment may be required for several weeks to months.

Candidiasis

This affects moist skin areas (groin, axilla, skin folds, nailfolds) or damaged skin (e.g. irritant dermatitis in nappy area of babies). It causes erythema with vesicles and pustules. Treatment is with topical or oral azole antifungals (e.g. miconazole, fluconazole).

Pityriasis versicolor

This is a mild skin infection caused by the dimorphic fungus, *Malassezia furfur*. It is found worldwide, but is most common in the tropics. Dark,

Table 44.1 Common dermatophyte (ringworm) infections.

COMMON DERMATOPHYTE INFECTIONS			
Infection	Site	Dermatophyte	Clinical features
Tinea corporis	Trunk	*E. floccosum*	Pruritus; circular erythematous lesion, scaling, clearing from centre
	Limbs	*Trichophyton* spp. *Microsporum* spp.	
Tinea cruris	Groin	*E. floccosum* *Trichophyton rubrum*	Pruritus; lesions similar to tinea corporis
Tinea pedis	Feet	*E. floccosum* *Trichophyton* spp.	Pruritus; cracking and scaling between toes
Tinea unguium	Nails	*Trichophyton* spp.	Thick, yellowish nails with surrounding erythema
Tinea capitis	Hair	*Trichophyton tonsurans*	Erythematous patches on scalp with scaling; hairs break, leaving bald patches
		Microsporum audounii *Microsporum canis*	

scaling lesions appear, particularly on the upper body. In dark-skinned patients, the affected areas become depigmented. Treatment is with topical or oral azole antifungals.

Infections of the eye

Eyelid

Eyelash follicle infections (styes) are usually caused by *S. aureus*.

Orbital cellulitis

Cellulitis of periorbital tissues caused principally by β-haemolytic streptococci, *S. aureus* and, rarely, *H. influenzae* type b in children (< 5 years).

Conjunctivitis

- *Viral conjunctivitis*: aetiology includes adenoviruses and measles virus. Laboratory diagnosis is by viral culture. There is no specific treatment.
- *Neonatal conjunctivitis* (ophthalmia neonatorum) is described in Chapter 46.
- *Bacterial causes* include *S. aureus*, *H. influenzae* and pneumococci. It is most common in young infants. Treatment with topical antibiotics is normally effective.

Herpesvirus infections of the eye

- *Herpes simplex virus*: conjunctivitis and painful corneal ('dendritic') ulcers occur; these may result in corneal scarring. Latency with recurrent infection may occur. Treatment is with aciclovir ointment.
- *Varicella-zoster virus*: ophthalmic zoster; the virus is latent in ophthalmic ganglia; it reactivates to affect the periorbital area (unilateral). Treatment is with systemic aciclovir.
- *Cytomegalovirus*: choroidoretinitis; associated with AIDS. Treatment is with ganciclovir.

Trachoma

Chlamydia trachomatis infection of the eye (conjunctivitis and corneal lesions) is prevalent in developing countries in the tropics, where it is an important cause of blindness. Spread is by direct hand–eye contact. Treatment is with tetracycline. Mass treatment campaigns have helped to eradicate trachoma in some areas where trachoma was endemic.

Other eye infections

- *Pseudomonas aeruginosa* may cause serious deep

Table 44.2 Common viral skin rashes.

Disease	Virus	Rash distribution	Associated symptoms and signs	Diagnosis	Treatment
COMMON VIRAL SKIN RASHES					
Maculopapular rashes					
Measles	Measles	Start from hairline downwards all over the body	Fever, coryza, conjunctivitis, Koplik's spots	Serology (IgM), antigen detection in respiratory specimens, culture	
Rubella	Rubella	Start from hairline downwards all over the body	Fever, lymphadenopathy (suboccipital, periauricular and cervical, arthritis in adults)	Serology (IgM), culture	
Erythema infectiosum (fifth disease)	Parvovirus B19	Starts on cheeks (slapped cheek) then diffuses a lace-like rash on body	Fever, arthritis in adults	Serology (IgM), polymerase chain reaction (PCR)	
Exanthem subitum (roseola infantum)	Human herpesvirus 6 (HHV6)	Face, trunk	2–3 days of fever disappearing with appearance of rash	Serology, PCR	Generally asymptomatic but ganciclovir or foscarnet in complicated cases
Enteroviruses infections	Several Coxsackieviruses and echoviruses	Scattered all over the body	Fever, myalgia, headache, cervical lymphadenopathy	Culture	

Table 44.2 *continued*

COMMON VIRAL SKIN RASHES

Disease	Virus	Rash distribution	Associated symptoms and signs	Diagnosis	Treatment
Primary HIV infection	HIV	Scattered all over the body	Glandular fever-like illness, meningitis, encephalitis	Serology, PCR	HAART
Dengue fever	Dengue virus	Scattered all over body	Fever, myalgia, headache PCR	Serology (IgM), culture,	
Vesicular rashes					
Chickenpox	Varicella-zoster virus (VZV)	Centripetal distribution	Fever, itching	Electron microscopy, antigen detection in vesicular fluid, serology (IgM), PCR	Generally symptomatic but aciclovir, valaciclovir or famciclovir in complicated cases
Shingles (Plate 68)	VZV	Dermatomal but could be disseminated	Post-herpetic neuralgia	Electron microscopy, antigen detection in vesicular fluid, serology (IgM), PCR	Aciclovir, valaciclovir or famciclovir
Disseminated herpes simplex	Herpes simplex virus (HSV)	Wide distribution	Neonatal herpes, highly immunosuppressed patients	Electron microscopy, antigen detection in vesicular fluid, PCR	Aciclovir, foscarnet
Smallpox	Smallpox	Centrifugal distribution	Fever, headache	Electron microscopy, PCR	
Hand foot and mouth disease	Enteroviruses (Coxsackievirus A16, A4, A5, A9, B2, B5)	Hands and feet	Mouth ulcers, fever	Culture	

eye infections after trauma or surgery. Treatment includes parenteral antibiotics.

- *Toxoplasma gondii* causes choroidoretinitis, normally as part of congenital toxoplasmosis.
- *Toxocara canis* causes choroidoretinitis.
- *Onchocerciasis* is a nematode infection. Micro-filariae invade the anterior chamber of the eye. It is an important cause of blindness in endemic areas (west Africa, Central America).
- *Acanthamoeba* sp. (Plate 39) results in an amoebic infection causing keratitis. It is associated with contact lens use.

Chapter 45

Infections in the compromised host

The immune system consists of adaptive (cellular and humoral) and innate mechanical barriers—skin and mucous membranes, complement and phagocytic cell mechanisms (see Chapter 23). When any of these protective mechanisms is reduced the host is considered to be immunocompromised. Such patients are more liable to have severe infections and can become infected with either a conventional pathogen (which can also cause disease in a non-compromised individual) or opportunistic pathogens (organisms that are usually unable to cause a disease in a healthy person). Various types of defect in the immune defence mechanisms can predispose to infections with certain pathogens.

Infections associated with defects in innate immune systems

(Table 45.1)

Infections associated with prosthetic devices

• *Intravenous and peritoneal dialysis catheters*: these catheters breach the skin barrier, allowing commensal organisms access to the deeper sites. The infections are commonly associated with coagulase-negative staphylococci, *Staphylococcus aureus*, and, less frequently, Gram-negative aerobic bacilli and *Candida* spp.
• *Prosthetic valves and joints*: coagulase-negative

staphylococci are the most common pathogens, gaining access usually during surgery. The microorganism can attach to the prosthesis in biofilms, acting as a nidus of infection.

Infections associated with burns

Burns can destroy extensive areas of the mechanical barriers of the body and also result in abnormalities in localized neutrophil function and antibody response. The burn exudates produce nutrition for microorganisms to colonize and cause infection. The usual pathogens include: *Pseudomonas aeruginosa*, *S. aureus*, *Streptococcus pyogenes*.

Infections associated with deficiencies in normal clearance mechanisms

• *Urinary catheters* allow microorganisms access to the bladder, overcoming defences, including urine washout. Bacteria can grow up the inside of a urinary catheter. The infections are commonly associated with Gram-negative aerobic bacilli from the patient's own faecal or periurethral flora.
• Stones in the kidney, common bile and salivary ducts can result in infections proximal to the obstruction.
• *Respiratory tract*: there may be damage to the cilia, e.g. with cystic fibrosis patients, which predisposes to pneumonia with organisms such as *S. aureus*, *Haemophilus influenzae* and *P. aeruginosa*.

Table 45.1 Infections associated with defects in mechanical barriers.

INFECTIONS	
Defect/condition	**Common pathogen**
Intravascular and peritoneal dialysis catheters	*Staphylococcus epidermidis* (most common), *S. aureus* coliforms; *Candida* spp.
Urinary catheter	Coliforms
Prosthetic valves and joints	*S. epidermidis*
Burns	Coliforms, *Pseudomonas*, *S. aureus* *Streptococcus pyogenes*
Biliary obstruction	Coliforms, *Enterococcus* spp.
Trauma/surgery	*S. aureus*

Table 45.2 Factors affecting adaptive systems.

FACTORS AFFECTING ADAPTIVE IMMUNITY		
Type	**Defect (example)**	**Notes**
Primary	B- and T-cell combined immunodeficiency	
Secondary	Chemotherapy	For example, cyclophosphamide
	Infections	For example, HIV-1 and -2
	Malignancy	For example, leukaemia, lymphoma
	Splenectomy	Reduced levels of immunoglobulin subclasses
	Irradiation	Affects proliferation of lymphoid cells
	Corticosteroids	

Table 45.3 Examples of infections related to deficiency in cellular and humoral immunodeficiency.

INFECTIONS		
Defensive defect	**Examples of underlying condition**	**Opportunistic organism**
T-cell	Lymphoma	Varicella-zoster virus
		Candida
		Toxoplasma gondii
T- and B-cell	Chronic lymphatic leukaemia	Varicella-zoster virus
		Candida
	Immunosuppressive drugs	*Toxoplasma gondii*
		Cytomegalovirus
		Pneumocystis jirovecii (P. carinii)
Neutropenia	Acute leukaemia	*Staphylococcus aureus*
	Cytotoxic treatment	*Aspergillus fumigatus*
		Coliforms
		Cytomegalovirus
		Pneumocystis jirovecii (P. carinii)

Table 45.4 Opportunistic pathogens associated with solid organ transplant recipients.

OPPORTUNISTIC PATHOGENS				
Bacteria	**Fungi**	**Protozoa**	**Parasites**	**Viruses**
Staphylococcus aureus	*Candida* spp.	*Toxoplasma* sp.	*Strongyloides stercoralis*	Herpes simplex viruses
Coagulase-negative staphylococci	*Aspergillus* spp.			Cytomegalovirus
Mycobacterium avium-intracellulare	*Pneumocystis jirovecii (P. carinii)*			Varicella-zoster virus
Enterobacteriaceae				Epstein–Barr virus

Infections associated with defects in complement and phagocytic activity

These defects are normally congenital. Defects in phagocytic function may affect chemotaxis, phagocytosis or bacterial killing; associated infections often involve pyogenic bacteria, e.g. *S. aureus*. Defects in the complement system are associated with bacterial infections, e.g. patients with defects in the later complement components (C7–C9) have an increased risk of meningococcal infections.

Infections related to cellular and humoral immunodeficiency

The type of underlying immunodeficiency (Table 45.2) may be related to the nature and severity of any subsequent infection (Table 45.3). Increas- ingly, these infections are iatrogenic and caused by opportunistic pathogens. The length of time that a patient is immunosuppressed may influence the type of associated infection.

Infections in some immunocompromised pa- tients can be multifactorial and associated with several factors, e.g. in recipients of solid organ transplants these are: the type of organ transplant; the immune suppression required to prevent rejec- tion of a grafted organ; and the pathogens to which a recipient is exposed, including from donor organ and the environment. Some infections asso- ciated with organ transplant recipients are shown in Table 45.4.

Generally conventional pathogens cause infec- tions in these patients before the level of immuno- suppression increases when more opportunistic infections start to appear.

Chapter 46

Perinatal and congenital infections

Definitions

Although congenital and neonatal infections are often described separately, there is considerable overlap in aetiology and pathogenesis.

Congenital infections

- These infections are acquired *in utero*, via either the placenta or genital tract (ascending infection).
- In the first trimester they are normally acquired transplacentally and cause developmental defects (congenital malformations); may result in spontaneous abortion, or a neonate born with mild or severe defects.
- If acquired later in pregnancy they may result in stillbirth or a neonate with an active infection.

Neonatal infections

- Infection during the neonatal period may be acquired *in utero* or from the genital tract during delivery.
- Typically present early in the neonatal period.
- Late-onset neonatal infections may be acquired from attending adults (e.g. parents, nursing staff), other children or the hospital environment.

Congenital infections (Table 46.1)

Diagnosis

This may be difficult with some intrauterine infections as maternal infection may be asymptomatic and clinical manifestations in the infant may appear only months or years after birth. The important microbiological investigations are:

- confirmation of infection in the mother during pregnancy, principally by serological tests
- isolation of the organism from the fetus or placenta, or from samples taken from the neonate (e.g. blood, faeces, urine, cerebrospinal fluid [CSF])
- serological tests on neonatal blood.

The presence of raised levels of specific IgG class antibodies (especially in the 6 months after birth) may reflect merely the presence of maternal antibodies, because IgG antibodies can cross the placenta. However, the presence of specific IgM class antibodies is diagnostic of intrauterine acquired infection, because maternal IgM antibodies cannot cross the placenta.

Rubella virus

Epidemiology

The risk of developing congenital infection after maternal rubella virus infection depends on the stage of pregnancy in which the maternal infection occurs; multiple severe defects are most common when infection occurs during the first 8

Table 46.1 Important congenital conditions.

IMPORTANT CONGENITAL CONDITIONS	
Organism	**Congenital malformation**
Rubella virus	Congenital heart defects; cataracts; microcephaly; deafness; hepatosplenomegaly, purpura
Cytomegalovirus	Deafness; microcephaly; low birthweight
Varicella-zoster virus	Low birthweight; encephalitis, limb hypoplasia; neonatal chickenpox (severe and widespread)
Human immunodeficiency virus	Paediatric AIDS
Treponema pallidum	Hepatosplenomegaly; bone and tooth defects; microcephaly
Toxoplasma gondii	Choroidoretinitis; hydrocephalus; microcephaly
Listeria monocytogenes	Intrauterine death; neonatal infection (meningitis)

weeks of pregnancy. Since the introduction of widespread or universal rubella immunization, the incidence of congenital rubella syndrome has decreased.

Clinical features

Rubella affects blood vessels in developing organs, causing: congenital heart defects, eye defects (cataracts, retinopathy, microophthalmos), deafness, microcephaly, seizures and learning disorders, hepatosplenomegaly and thrombocytopenia.

Laboratory diagnosis

By the presence of rubella-specific IgM antibodies in the mother during early pregnancy, isolation of virus from the urine or throat of an infected neonate and the presence of rubella-specific IgM antibodies in neonatal blood.

Prevention

Immunization of girls and identification of non-immune women by antenatal screening followed by immunization postpartum (the vaccine contains live virus and therefore cannot be given during pregnancy) have led to a significant reduction in both the number of rubella-susceptible women and congenital rubella in developed countries. In the UK, boys are also immunized in an attempt to lower the pool of rubella-susceptible individuals in the community even further.

Cytomegalovirus (CMV)

Epidemiology

• Given the successful immunization strategy against rubella virus infection, congenital infection caused by CMV is now the most common intrauterine infection (1% of neonates in the UK and the USA).

• CMV intrauterine infection may follow primary CMV infection or, less commonly, reactivation of latent infection (secondary infection). Fetal damage may occur after maternal infection in any trimester, but is more serious when infection occurs early in pregnancy; about 40% of fetuses become infected after primary infection during pregnancy, *but* 90% will be normal at birth and only a small percentage of these will later present with deafness and/or learning disorders.

Clinical features

These include pneumonitis, hepatosplenomegaly, thrombocytopenia, anaemia; low birthweight, microcephaly, cerebral palsy, deafness and choroidoretinitis.

Laboratory diagnosis

• Maternal CMV infection in pregnancy is diagnosed by detection of CMV-specific IgM.

• Neonatal infection is confirmed by the detection of CMV-specific IgM and the isolation of the virus or the detection of viral DNA by molecular techniques in throat swabs or urine. Excretion of CMV in the urine may continue for several months, and babies with congenital CMV infection should be isolated while in hospital.

Treatment and prevention

The strategy followed for congenital rubella prevention is not possible because there is currently no vaccine available for CMV. For this reason, and as severe disease after intrauterine infection is rare, antenatal screening is not performed routinely. Ganciclovir is recommended for the treatment of congenitally infected babies with evidence of central nervous system (CNS) infection.

Toxoplasmosis

Epidemiology (see Chapter 21)

Maternal infection is often subclinical or presents as a mild influenza-like illness. In the UK toxoplasmosis affects about 0.5% of pregnancies. Fetal damage most frequently occurs in the first trimester.

Clinical features

Stillbirth, choroidoretinitis, microcephaly, convulsions, intracerebral calcification with resultant hydrocephalus, hepatosplenomegaly and thrombocytopenia.

Laboratory diagnosis

Detection of *Toxoplasma*-specific IgM antibodies in the mother during pregnancy or in the neonate.

Treatment and prevention

Spiramycin, pyrimethamine and sulphadiazine treatment during pregnancy may reduce the severity of sequelae. There is no vaccine for *Toxoplasma*, but pregnant women are advised to avoid contact with cat faeces during pregnancy. Antenatal screening for toxoplasma antibodies is performed in some countries.

Varicella-zoster virus (VZV)

Maternal chickenpox between 7 days before and 7 days after delivery may result in severe neonatal infection; in such circumstances, neonates should be given specific immunoglobulin to varicella-zoster (ZIG) and/or aciclovir immediately after birth. Rarely (< 1%), maternal infection in the first 20 weeks of pregnancy (highest risk between 12 and 20 weeks) may result in congenital defects (low birthweight, encephalitis, choroidoretinitis, hyperplastic limbs). Non-immune pregnant women should be advised to avoid contact with cases of chickenpox; in cases of proven contact in seronegative pregnant women, they should be treated with ZIG.

Parvovirus

It is estimated that 30% of maternal parvovirus infections lead to fetal infection. The highest risk period is the first 20 weeks of pregnancy. Maternal parvovirus infection results in spontaneous abortion in about 10% of cases. The virus infects erythrocyte precursors, resulting in anaemia and cardiac failure in the fetus; congenital malformations have not been reported. Maternal infection with parvovirus can be confirmed by the detection of a specific IgM. There is no vaccine to parvovirus. Pregnant women should be advised to avoid contact with children with suspected parvovirus infection.

Syphilis

Epidemiology

Congenital syphilis is now rare in developed countries because of routine antenatal screening, but remains a serious problem in developing countries. Primary or secondary syphilis occurring during pregnancy may result in intrauterine infection.

Clinical features

Stillbirth, hepatosplenomegaly, lymphadenopathy, skin rashes, nasal discharge, bone, tooth and cartilage defects (e.g. saddle-shaped nose), and deafness.

Laboratory diagnosis

By detection of IgM specific to *Treponema pallidum* in neonatal serum.

Prevention

There is routine antenatal screening for the presence of active syphilis in pregnant women; detection should be followed by prompt treatment with penicillin.

Tuberculosis (TB)

Intrauterine infection with TB remains a problem in areas of the world where the disease is common. Infection early in pregnancy may result in abortion; later in pregnancy it may result in the birth of a neonate with active TB, often presenting as pneumonia or hepatitis.

Listeria monocytogenes

Maternal infection (asymptomatic or influenza-like illness) may be acquired from soft cheeses, and raw vegetables or salads. It may result in intrauterine infection, with abortion or stillbirth. Infection late in pregnancy may result in the birth of an infected neonate with septicaemia and meningitis. Prevention is based on avoiding certain foods in pregnancy.

Malaria

Congenital or neonatal malaria occurs most frequently when the mother becomes infected for the first time during pregnancy. Infection may result in abortion or stillbirth or pre-term delivery. Neonates may not develop symptoms of malaria for several weeks after delivery.

Neonatal infections

Neonates, particularly when pre-term, are immunocompromised because of an immature immune system. Minor sepsis can lead rapidly to septicaemia and shock, with a high mortality. Advances in intensive care have led to increased numbers of pre-term neonates surviving. This has increased the use of long-term intubation, indwelling intravascular lines and other monitoring devices, and antibiotic treatment. Early diagnosis of infection is difficult; signs of sepsis may be absent or obscure, e.g. decreased alertness, temperature instability, reduced feeding.

Group B β-haemolytic streptococci

A commensal of the adult genital tract and perianal area; about 30–50% of pregnant women are colonized. Subsequent colonization of newborn babies is common, but only a small number become infected; pre-term and low-birthweight infants are particularly at risk. Infections may present early (less than 48h after delivery) or late (1 week to 6 months after delivery). Typically, early onset infections are more severe, with pneumonia and septicaemia; late-onset infections often present as meningitis. Treatment is penicillin with gentamicin. Prevention is by using intrapartum antibiotics for those women who are identified as carrying group B streptococci.

Listeria monocytogenes

This is acquired perinatally, probably from the maternal intestinal tract. It is an important cause of neonatal meningitis and septicaemia during the first few weeks of life and also a cause of septic abortion. Treatment is with ampicillin or penicillin.

E. coli

E. coli is an important cause of neonatal meningitis and septicaemia; acquisition is probably from the maternal intestinal tract. Invasive infections are associated with certain *E. coli* serotypes, particularly the K1 serotype. Treatment is various, according to antibiotic susceptibility tests.

Other Gram-negative aerobic bacilli

Klebsiella spp., *Pseudomonas aeruginosa* and other antibiotic-resistant Gram-negative bacilli can be acquired from the hospital environment and are often associated with prolonged hospitalization in special-care baby units. Infections include pneumonia, septicaemia, meningitis and urinary tract infections. Treatment is often with antibiotic combinations, guided by antibiotic sensitivity tests.

Coagulase-negative staphylococci

The increased use of indwelling intravascular catheters during the neonatal period has led to an increase in the number of associated infections caused by coagulase-negative staphylococci, particularly *Staphylococcus epidermidis*. This is now the most common organism isolated from neonatal blood cultures. Management includes possible removal of infected catheters and the use of antibiotics such as vancomycin or teicoplanin.

Staphylococcus aureus

This organism may cause pneumonia, intravascular catheter infections, skin infections (e.g. scalded skin syndrome) and conjunctivitis. Treatment is usually with flucloxacillin (if meticillin-resistant *S. aureus* (MRSA), use vancomycin).

Neonatal eye infections

Aetiology and epidemiology
Neisseria gonorrhoeae and *Chlamydia trachomatis* are acquired from the maternal genital tract; maternal infection may be subclinical. Other causes include *S. aureus, Haemophilus influenzae* and pneumococci that are acquired from the infant's own flora or from close contacts.

Clinical features
• *N. gonorrhoeae*: severe conjunctivitis with pus and periorbital inflammation; frequently presents within 48 h of birth.
• *C. trachomatis*: less severe than gonococcal conjunctivitis; often presents 3–7 days after delivery; if untreated, chlamydial pneumonitis may occur as a late complication (1–3 months); associated with pneumonia.
• *S. aureus, H. influenzae, S. pneumoniae*: purulent conjunctivitis; normally presents 2–5 days after delivery.

Laboratory diagnosis
• *N. gonorrhoeae*: Gram stain (intracellular Gram-negative diplococci) and culture of conjunctival swab; appropriate investigations for maternal infection, including cervical swabs.
• *Chlamydia* spp.: culture of conjunctival swab or direct detection of antigen by fluorescence or enzyme-linked immunosorbent assay (ELISA) or molecular techniques; appropriate investigations for maternal infection, including cervical swabs.
• *S. aureus, H. influenzae, S. pneumoniae*: conjunctival swabs for routine bacterial culture.

Treatment and prevention
• *N. gonorrhoeae*: prevention is by antenatal screening and treatment of pregnant women with gonorrhoea. Treatment to the infant, e.g. penicillin or ceftriaxone, is based on the local antibiotic sensitivities. The routine administration of parenteral penicillin or antibiotic eyedrops (chloramphenicol) to all newborn babies is carried out in areas of the world with a high incidence of gonorrhoea.
• *C. trachomatis*: tetracycline eye ointment plus oral erythromycin.
• *S. aureus, H. influenzae, S. pneumoniae*: antibiotics (e.g. chloramphenicol), eyedrops or ointment; systemic antibiotics are sometimes required (e.g. flucloxacillin for *S. aureus*).

Herpes simplex virus (HSV)

Epidemiology
Acquired from the maternal genital tract during delivery; asymptomatic shedding of HSV occurs in about 2–5% of pregnant women; the incidence is higher in women of lower socioeconomic groups. Maternal infection may be primary or recurrent. The risk of acquisition is increased in maternal primary infections because excretion of the virus is higher than in recurrent infection.

Clinical features
Encephalitis or meningoencephalitis, generalized infection with hepatitis, and conjunctivitis.

Laboratory diagnosis
This is by direct immunofluorescence of vesicular

fluid from skin lesions, or culture of virus or detection of viral DNA from throat washings, cerebrospinal fluid (CSF) or conjunctival swabs.

Treatment and prevention

Treatment of herpes encephalitis with high doses of aciclovir has reduced the mortality rate to around 25%. Symptomatic genital herpes at term is an indication for caesarean section.

Preventive strategies based on routine antenatal screening for genital herpes are impractical because virus shedding is intermittent and the cost would be high; screening of high-risk groups has been advocated in some countries.

Human immunodeficiency virus (HIV)

Epidemiology

HIV transmission from infected mothers to their babies can occur *in utero* (< 1% of cases) during delivery, or postpartum via breast-feeding. The risk of transmission has been estimated to be 15–40%, if no preventive measures were taken. The risk is higher if the mother has advanced disease with a low CD4 cell count and a high HIV viral load, with prolonged rupture of membrane during labour, vaginal delivery and breast-feeding.

Clinical features

Poor weight gain, lymphadenopathy, hepatosplenomegaly and an increased susceptibility to infection.

Laboratory diagnosis

A definitive diagnosis can only be confirmed by the detection in the neonate of HIV RNA or proviral DNA by gene amplification techniques (polymerase chain reaction [PCR]). Infants are normally screened within 2 weeks, 6 weeks, 3 months and 6 months of birth. Persistence of HIV antibody in the infant for more than 18 months after birth is also diagnostic.

Prevention

Screen pregnant women for HIV. Treat infected mother with antiretroviral drugs before, during and after labour; caesarean section; treat infant with antiretroviral drugs.; avoid breast-feeding if possible.

Hepatitis B virus (HBV)

Epidemiology

Vertical transmission of HBV is the most important route of infection; worldwide, vertical transmission is the most frequent route, particularly in high endemic areas. Acquisition rates vary according to hepatitis B 'e' antigen status of the mother (70–90% if the mother is HBeAg positive and 10–40% if HBeAg negative). Transmission occurs mostly at the time of delivery.

Clinical features

Infection is often subclinical, but 90% of infected infants will become chronic carriers with subsequent cirrhosis and are at a high risk of developing liver cancer in later life.

Laboratory diagnosis

This is dependent on detection of hepatitis B surface antigen (HBsAg) in the infant's blood, which persists for months.

Prevention

Administration of specific HBV immunoglobulin as early as possible after delivery and simultaneous active immunization with hepatitis B vaccine (given at birth, and 1 and 2 months after birth).

Enteroviruses

Neonatal sepsis caused by enterovirus infection is a serious disease with high mortality rates (> 10%). Infected neonates are at greatest risk when signs and symptoms develop in the first few days of life. Infected babies will present with fever, vomiting, maculopapular rash and signs of meningeal inflammation. The disease may progress with the infants developing hepatic necrosis, myocarditis and encephalitis. Diagnosis is by isolation of viruses or detection of viral RNA by molecular techniques in throat swabs, nasopharyngeal aspirate (NPA), stools or CSF. Treatment is mainly symptomatic.

Obstetric infections

Puerperal sepsis

This is a uterine infection after delivery or termination of pregnancy, and is now uncommon in developed countries. Important pathogens include *S. pyogenes*, *Clostridium perfringens*, *E. coli* and *Bacteroides* spp. Predisposing factors include: instrumentation, premature rupture of membranes, retained placental material, and the use of non-sterile equipment/poor practice, e.g. illegal abortion. Laboratory investigations include culture of high vaginal swab and blood cultures. Treatment is with combinations of antibiotics to cover likely pathogens (e.g. penicillin, gentamicin plus metronidazole).

Chapter 47

Human immunodeficiency viruses

The viruses

Human immunodeficiency virus (HIV) types 1 and 2 are members of the Retroviridae family of viruses, subfamily Lentiviruses. Retroviruses have a diploid RNA genome and encode RNA-directed DNA polymerase (reverse transcriptase [RT]). When cells become infected with HIV, the RT enzyme catalyses the synthesis of double-stranded DNA that subsequently becomes integrated into host chromosomal DNA. This integrated viral DNA serves as a template for viral genomic and messenger RNA transcription by the host cell's synthetic and processing system.

HIV-1 is divided into three groups: the main group (M), the new group (N) and the outlier group (O). Groups N and O remain largely confined to a part of west central Africa (Gabon and the Cameroon), although sporadic infection through contact with people from that region occurs. Group M viruses spread widely to cause the worldwide AIDS (acquired immune deficiency syndrome) pandemic. Group M viruses are further divided into subtypes (clades) lettered from A to K.

The viruses measure approximately 100–150nm in diameter. Mature viral particles are characterized by an electron-dense cylindrical core surrounded by a lipid envelope acquired as the virion buds from infected cells. The virion core is composed of the matrix (p17), capsid (p24) and nucleocapsid

(p7) proteins enclosing the viral genomic RNA and polymerase enzymes. The viral envelope contains the surface (gp120) and transmembrane (gp41) glycoproteins.

The HIV genome carries three major genes: *gag*, *pol* and *env*: *gag* encodes the proteins that form the structural core and matrix of the virus; *pol* encodes the RT, protease and integrase enzymes necessary for virus replication; and *env* encodes the envelope proteins. The genome also includes several other regulatory and accessory genes.

Epidemiology

Modes of transmission

- Sexual (in the western world mainly homosexual transmission, whereas across the world it is heterosexual transmission)
- Intravenous drug use
- Contaminated blood and blood products (less of a burden after the introduction of donation screening for HIV)
- Perinatal transmission from mother to child and through breast-feeding.

Global picture at the end of 2005

- 63 million people are estimated to be infected worldwide, 42 million living with HIV and 23 million deaths

- 29 million infected in Africa (in some countries in sub-Saharan Africa, one in four people in the population is infected)
- 5 million new cases in 2005
- 3.2 million deaths in 2005.

Subtype distribution

In Africa subtypes A, C, D and F predominate, whereas subtype B is the most common subtype in the western world.

Pathogenesis

HIV infects mainly CD4+ cells by binding to the CD4 molecule as a receptor. CD4+ cells include lymphocytes, monocytes, macrophages, dendritic cells and macroglial cells. Other molecules (co-receptors) are required for HIV entry into cells. These molecules include the chemokine receptors CCR5 and CXCR4. After primary infection and seroconversion, there follows an asymptomatic phase of infection lasting 2–15 years. During this period high turnover of HIV production and lymphocyte replenishment occurs. Infected lymphocytes are depleted through direct cytopathic effect of HIV, by cytotoxic T lymphocytes and through apoptosis.

Clinical features

Primary infection

A mononucleosis-like seroconversion illness occurs in 50–90% of infected patients. The syndrome is characterized by fever, myalgia, maculopapular rash, sore throat, cervical lymphadenopathy, aseptic meningitis and, in rare cases, encephalitis. There is a high plasma viral load and CD4 lymphopenia. Primary infection usually lasts less than 14 days and is self limiting.

The asymptomatic phase

The duration of this phase varies between 2 and 15 years. Lymphadenopathy may be a complaint in some cases.

Symptomatic disease and AIDS

In untreated patients, this phase is characterized by opportunistic infections and malignancies (Table 47.1) caused by severe immunodeficiency. Patients complain of weight loss, night sweats, recurrent chest infections and oral candidiasis.

Diagnosis and follow-up

HIV antibody detection

Several techniques are available, e.g. enzyme immunoassay (EIA), particle agglutination and western blots. Specimens reactive in one assay format should be confirmed by at least two other assays.

Detection of p24 antigen

By EIA in early and late infection.

Nucleic acid detection

For example, RT polymerase chain reaction (PCR) for diagnosis of infection and quantitative measurement of viral load in untreated and treated patients. This helps to determine disease stage, to monitor progression, to assess response to antiretroviral therapy and to provide evidence of early treatment failure.

Antiviral resistance testing

Phenotypic (drug-sensitive culture assays) and genotypic (genome sequencing to detect mutation associated with resistance) assays are available for monitoring response to antiretroviral treatment.

Treatment

Several classes of antiretroviral drugs are available for HIV treatment (see Chapter 31). A minimum of three drugs should be used (highly active antiretroviral therapy [HAART]). Start of treatment is generally guided by viral load measurement and CD4 count. Monitoring of the response is aided by viral

Table 47.1 Opportunistic infections and malignancies during AIDS.

OPPORTUNISTIC INFECTIONS AND MALIGNANCIES DURING AIDS	
Bacterial infections	*Mycobacterium avium-intracellulare* (disseminated)
	M. kansasii (disseminated)
	M. tuberculosis (disseminated or pulmonary)
	Recurrent disseminated salmonellosis
	Recurrent bacterial pneumonia
Viral infections	CMV
	HSV—persistence of lesions for > 1 month
Fungal infections	*Pneumocystis jirovecii* (*carinii*) pneumonia
	Candidiasis
	Cryptococcosis (extrapulmonary)
	Coccidioidomycosis
Protozoal infection	Cryptosporidiosis
	Toxoplasmosis
	Histoplasmosis
	Strongyloidosis
Malignancies	Kaposi's sarcoma
	Non-Hodgkin's lymphoma
	Primary CNS lymphoma
	Cervical carcinoma

load and testing for evidence of resistance (see Chapter 41).

Prevention

• Public health measures, e.g. 'safe sex' education, needle exchange programmes.

• Screening of blood, blood products and organ donors.

• Prevention of perinatal transmission by:
  — treatment of mothers and newborn babies with antiretroviral drugs
  — caesarean sections
  — avoidance of breast-feeding if possible.

Chapter 48

Miscellaneous viral infections

Arboviruses

Arboviruses are maintained in nature principally or to an important extent through biological transmission between susceptible vertebrate hosts by haematophagous arthropods. Natural vertebrate hosts include birds and rodents, but transmission may also occur to domestic animals and humans. The disease spectrum in the vertebrates varies from clinically unapparent infection to severe disease and death. Prevention of transmission is by vector control and avoidance of insect bites (clothing, DEET insecticide, 'mosquito' netting when sleeping).

All medically important arboviruses are members of three families of viruses: *Togaviridae*, *Flaviviridae* and *Bunyaviridae*.

Yellow fever

Aetiology and epidemiology
Genus flavivirus. Transmitted by mosquitoes, especially *Aedes aegypti*. Major vertebrate host is the monkey. Three transmission cycles:

1 Enzootic forest cycle between mosquitoes and monkeys. Human rarely infected when venturing into the forest.
2 Jungle yellow fever cycle. Most important cycle in respect to human infection when forest mosquitoes invade adjacent plantations, clearings and villages. Once infection has been introduced to the human host, human-to-human transmission sustains the epidemic.
3 Urban yellow fever cycle. Transmission between humans maintained by mosquitoes.

The disease occurs in the tropics on both sides of the Atlantic. Control of the mosquitoes in the Americas resulted in the disappearance of urban yellow fever from the western hemisphere, but an average of 100 cases/year of jungle yellow fever are still reported. In Africa, epidemics of the disease continue to occur regularly with high morbidity and mortality.

Clinical features
Incubation period is 3–6 days. There are two phases of the disease; an initial phase of fever, headache, muscle pain and nausea. Most cases are aborted at this stage. Second and severe phase starts after a few days of remission by the onset of haemorrhages (i.e. haematemesis, melaena), hepatic and renal disease. The mortality rate is 20–50%.

Diagnosis
Virus isolation in specialist reference laboratories using mosquito cell lines or suckling mice. Serology (IgG and IgM antibodies) and polymerase chain reaction (PCR).

Treatment and prevention
Symptomatic treatment. A live attenuated vaccine

(17D vaccine) is available for travellers to endemic areas and for use in response to outbreaks.

Dengue fever

The most important arboviral cause of disease and death in humans.

Aetiology and epidemiology

Genus flavivirus; four serotypes: 1–4. Transmitted by mosquitoes, especially *Aedes aegypti*. The only vertebrate hosts are humans and some species of Asian and African primates. Three cycles of transmission are recognized: forest, jungle and urban cycles (see 'Yellow fever'). The disease is endemic in subtropical and tropical regions of the world.

Clinical features

Majority of infections are asymptomatic. Clinical disease characterized by fever, headache, nausea and vomiting, severe back pain and wide spread maculopapular rash.

Complication

Dengue haemorrhagic fever and dengue shock syndrome in 2% of cases.

Diagnosis

Serology (IgG and IgM antibodies), virus isolation in mosquito cell lines and suckling mice, PCR.

Treatment and prevention

Symptomatic treatment. Prevention is by methods for vector control.

Japanese encephalitis

The major arboviral cause of encephalitis worldwide.

Aetiology and epidemiology

Genus flavivirus; three genotypes. Transmitted by mosquitoes, especially *Culex* spp. Vertebrate hosts are humans and domestic animals and birds. Humans are a dead-end host and play little role in the amplification of the virus. Endemic in south-east Asia.

Clinical features

Incubation period is 1–2 weeks. Infections may present as mild febrile illness, aseptic meningitis or meningoencephalitis.

Complications

Twenty-five per cent of encephalitis cases will recover with no permanent sequelae, 25% will die rapidly. The remaining 50% will recover with varying degrees of permanent neuropsychiatric sequelae.

Diagnosis

Serology (IgG and IgM antibodies) in serum and cerebrospinal fluid (CSF), virus isolation in mosquito cell lines and suckling mice, PCR.

Treatment and prevention

Treatment is symptomatic. A killed vaccine is wildly used in endemic areas and for travellers to endemic areas. A live attenuated vaccine is used in China. Vector control.

West Nile virus

Aetiology and epidemiology

Genus flavivirus. Transmitted by mosquitoes, especially *Culex* spp. Hosts include birds, domestic and wild animals, humans and subhuman primates. Humans are a dead-end host. No person-to-person transmission occurs but transmission through blood and organ donations and intrauterine transmissions has been reported. The virus is found throughout Africa, Asia, Europe, the USA and Canada.

Clinical features

The disease may present as mild febrile illness, aseptic meningitis or meningioencephalitis in 0.7% of patients.

Diagnosis

Serology (IgG and IgM antibodies) in serum and CSF, virus isolation in mosquito cell lines and suckling mice, PCR.

Table 48.1 Agents of viral haemorrhagic fevers.

AGENTS OF VIRAL HAEMORRHAGIC FEVERS	
Agent	Source
Arenaviridae	
Lassa fever	Rodents
Argentine haemorrhagic fever (Junin)	Rodents
Bolivian haemorrhagic fever (Machupo)	Rodents
Venezuelan haemorrhagic fever (Guanarito)	Rodents
Bunyaviridae	
Crimean–Congo haemorrhagic fever	Ticks
Haemorrhagic fever with renal syndrome (Hantaan viruses)	Rodents
Rift Valley fever	Mosquitoes
Filoviridae	
Ebola	Unknown
Marburg	Unknown
Falviviridae	
Dengue types 1–4	Mosquitoes
Yellow fever	Mosquitoes
Kyasanur Forest disease	Ticks
Omsk haemorrhagic fever	Ticks
Togaviridae	
Chikungunya	Mosquitoes

Treatment and control

Treatment is symptomatic. No vaccine available. Prevention is by vector control.

Viral haemorrhagic fever viruses

Viral haemorrhagic fevers are severe life-threatening diseases caused by a range of viruses (Table 48.1). Only four of theses viruses are known to be readily capable of person-to-person spread and therefore present a large public health risk. Diagnosis of these viruses is attempted only in biosafety level P4 laboratories. Strict infection control precautions, including patient isolation, are important to prevent transmission in hospitals.

Lassa fever

Aetiology and epidemiology

Member of the Arenaviridae family of viruses. The natural reservoir is the multimammate rat *Mastomys natalensis*. The virus produces a persistent tolerant infection in this host with no ill-effects. The rats remain infectious during their lifetime, freely excreting Lassa virus in urine and other body fluids. Humans become infected through contact with infected rodents or rodents' excreta. Person-to-person transmission can occur in overcrowded conditions and in particular in rural hospitals, where universal infection control precautions are suboptimal. Lassa is endemic in west Africa with several reported outbreaks mainly in Sierra Leone and Nigeria.

Clinical features

Incubation period is 3–16 days. Spectrum of disease ranges from subclinical to fulminating fatal infection. Symptomatic patients present with fever, sore throat, myalgia, abdominal pain and vomiting lasting between 7 and 17 days.

Complications

Haemorrhages, deafness, encephalitis and spontaneous abortion. Mortality rate is between 1 and 2%.

Diagnosis

Virus isolation in tissue culture, direct antigen detection by immunofluorescence in conjunctival cells, serology, PCR.

Treatment and prevention

Intravenous ribavirin within 6 days of onset of illness, Lassa immune plasma. No vaccine available.

Ebola and Marburg haemorrhagic fevers

Aetiology and epidemiology

Ebola and Marburg viruses are members of the Filoviridae family of viruses. No natural reservoir is identified and the natural history of these viruses is unknown. Viruses are endemic in Africa with several outbreaks in Sudan, Zaire, the Democratic Republic of Congo, Gabon, Uganda and Angola. During outbreaks humans are infected through close contacts with either infected patients or infected material such as needles, blood or secretions.

Clinical features

Incubation period is 3–9 days. Infections are always symptomatic, starting abruptly with fever, headache, myalgia, arthralgia, conjunctivitis, sore throat, abdominal pain, nausea and vomiting.

Complications

Haemorrhages, encephalitis. Mortality rate is between 55 and 88%.

Diagnosis

Virus isolation in tissue culture, serology, PCR.

Treatment and prevention

Treatment is symptomatic. No vaccine available.

Crimean–Congo haemorrhagic fever

Aetiology and epidemiology

Virus is member of the Bunyaviridae family of viruses. Vector is the ixodid tick. Hosts include humans, and domestic and wild animals. Humans become infected through tick bites or contact with blood, tissue or excreta of infected animals. Nosocomial transmission of the virus also occurs. The virus is distributed over eastern Europe, Asia and Africa.

Clinical features

Incubation period is 1–14 days. Sudden onset of fever, severe headache, sore throat, nausea and vomiting, general fatigue and malaise.

Complication

Bleeding tendency, i.e. epistaxis, haematuria, haematemesis, melaena. Mortality rate is 30%.

Diagnosis

Virus isolation in cell culture, serology (IgG and IgM antibodies), PCR.

Treatment and prevention

Treatment is generally symptomatic but the antiviral drug ribavirin has shown some benefit particularly if given within 5 days of onset of symptoms. No vaccine available.

Subject index

Page numbers in italic refer to figures or tables

Organism index

Page numbers in italic refer to figures or tables